T0134201

The Calculus Story

The Calculus Story

A Mathematical Adventure

DAVID ACHESON

OXFORD
UNIVERSITY PRESS

OXFORD
UNIVERSITY PRESS

Great Clarendon Street, Oxford, OX2 6DP,
United Kingdom

Oxford University Press is a department of the University of Oxford.
It furthers the University's objective of excellence in research, scholarship,
and education by publishing worldwide. Oxford is a registered trade mark of
Oxford University Press in the UK and in certain other countries

© David Acheson 2017

The moral rights of the author have been asserted

First Edition published in 2017

All rights reserved. No part of this publication may be reproduced, stored in
a retrieval system, or transmitted, in any form or by any means, without the
prior permission in writing of Oxford University Press, or as expressly permitted
by law, by licence or under terms agreed with the appropriate reprographics
rights organization. Enquiries concerning reproduction outside the scope of the
above should be sent to the Rights Department, Oxford University Press, at the
address above

You must not circulate this work in any other form
and you must impose this same condition on any acquirer

Published in the United States of America by Oxford University Press
198 Madison Avenue, New York, NY 10016, United States of America

British Library Cataloguing in Publication Data
Data available

Library of Congress Control Number: 2017935884

ISBN 978-0-19-880454-3

Printed in Great Britain by
Clays Ltd, Elcograf S.p.A.

Links to third party websites are provided by Oxford in good faith and
for information only. Oxford disclaims any responsibility for the materials
contained in any third party website referenced in this work.

In memory of
Dr Janet Mills
(1954–2007)

who once claimed that she never quite understood calculus

CONTENTS

1

Introduction

In the summer of 1666, Isaac Newton saw an apple fall in his garden, and promptly invented the theory of gravity.

That, at least, is the story.

And, however oversimplified this version of events may be, it makes as good a starting point as any for an introduction to calculus.

Because the apple *speeds up* as it falls.

1. Newton and the apple.

It even raises the whole question of what we mean, exactly, by the speed of the apple at any given moment.

This is because the well-known formula

$$speed = \frac{distance}{time}$$

only applies when the speed of motion is constant, i.e. when distance is proportional to time.

To put it another way, the formula only applies if the graph of distance against time is a straight line, the speed then being represented by the slope, or steepness of the line, as in Figure 2.

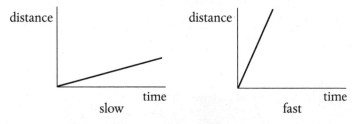

2. Motion at constant speed.

But, with a falling apple, distance isn't proportional to time. As Galileo discovered, the distance fallen in time t is proportional to t^2.

So, after a certain time the apple will have fallen a certain distance, but after twice as long it will have fallen not twice as far but four times as far, because $2^2 = 4$. And if we plot the distance fallen against time we get the curve in Figure 3, which bends upwards.

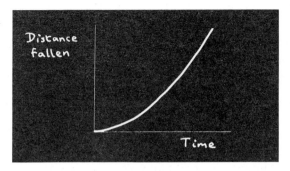

3. How an apple falls.

Plainly, the increasing steepness of the curve reflects, in some way, the increasing rate at which the apple falls, as time goes on.

And this idea of the rate at which something is changing with time is one of the most central ideas in the whole of calculus.

Calculus is sometimes said to be all about change, but a better description, arguably, is that it is all about the *rates* at which things change.

4. (a) Isaac Newton (1642–1727) (b) Gottfried Leibniz (1646–1716)

The subject came fully to life in the second half of the 17th century, largely through the work of Isaac Newton, in England, and Gottfried Leibniz, in Germany.

The two never met, but there was a certain amount of wary (and indirect) correspondence between them. At first, this was amicable and polite, but the relationship eventually deteriorated into a major row about who had 'invented' calculus.

While I will say more about this later, my main concern in this short book is with calculus itself.

Above all, I want to offer a 'big picture' of the subject as a whole, concentrating on the most important ideas, and something of their history.

We will see, also, how calculus is fundamental to physics and the other sciences.

One particular aim, for instance, will be to take the theory far enough that we can understand the vibrations of a guitar string (see Figure 5).

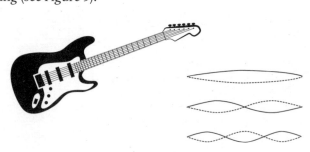

5. Guitar string vibrations.

But I will also stress, throughout the book, occasions on which results from calculus can be enjoyed purely for their own sake, regardless of any possible practical application.

Figure 6, for instance, shows an extraordinary connection between π—which is all about circles—and the odd numbers.

$$\frac{\pi}{4} = 1 - \frac{1}{3} + \frac{1}{5} - \frac{1}{7} + \cdots$$

6. A surprising connection.

And, in due course, I will try to show just why this result is true.

In short, then, this little book is more ambitious than it looks.

If all goes well, we will see not only what calculus is really about, *but how to actually start doing it.*

And to set about that, we need first to think a little about the very nature and spirit of mathematics itself.

2

The Spirit of Mathematics

In the Babylonian Collection at Yale University there is a famous clay tablet, known as YBC 7289. It dates from roughly 1700 BC, and has a simple geometrical figure on it (Figure 7).

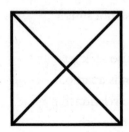

7. Square and diagonals.

The figure is accompanied by some cuneiform writing, and when that was deciphered it was found to be an approximation to the number $\sqrt{2}$—correct to better than 1 part in a million.

How, then, did the writer *know* that, for a square, the ratio of diagonal to side is $\sqrt{2}$?

We can only guess, I think, that they appealed to a diagram such as Figure 8.

The area of the large square is $2 \times 2 = 4$. The area of the shaded square is evidently half of this, and therefore 2. So the side of the shaded square must be $\sqrt{2}$.

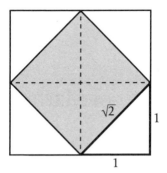

8. A simple deduction.

Today, this deductive aspect of mathematics is seen as central to the whole subject.

We continually ask not simply 'What is true?' but '*Why* is it true?'

Mathematicians also seek *generality* whenever possible, and Pythagoras' theorem is a famous example, for it provides an unexpectedly simple relationship between the three sides of *any* right-angled triangle – short and fat or long and thin.

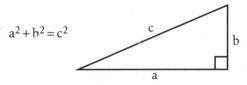

$$a^2 + b^2 = c^2$$

9. Pythagoras' theorem.

And, as with much that is best in mathematics, it is this generality which gives the theorem its power.

Algebra

While geometry dates back to ancient Greece and beyond, algebra—at least as we know it today—is a much more recent development.

Even the familiar equals '=' sign only appeared in 1557, less than a century before Newton was born.

The main purpose of algebra is, again, to help us express and manipulate general ideas in mathematics, in a succinct manner.

And one such result, of great value in this book, is

$$(x+a)^2 = x^2 + 2ax + a^2.$$

This is true for any numbers x and a, positive or negative, by the rules of elementary algebra, but when x and a are both positive it can even be seen geometrically, using areas (Figure 10).

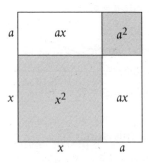

10. Algebra as geometry.

Proof

Sometimes in mathematics, the actual deductive arguments, or proofs, can be a source of pleasure in themselves.

Consider, for instance, the proof of Pythagoras' theorem in Figure 11.

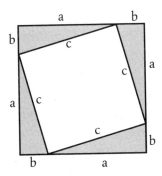

11. Proving Pythagoras' theorem.

Here, we have placed four copies of our right-angled triangle inside a square of side $a + b$, leaving a square of area c^2 in the middle.

Each right-angled triangle has area $\frac{1}{2} ab$, so the area of the large square is $c^2 + 2ab$.

But it is also $(a + b)^2 = a^2 + 2ab + b^2$.

So $a^2 + b^2 = c^2$.

I would argue that this is one of the best proofs of Pythagoras' theorem, in fact, because it is so concise and elegant.

The way to the stars...

Throughout its history, mathematics has played a crucial part in our understanding of how the world really works.

The nature of the Universe, in particular, has always been a source of wonder. Yet to study it, we must begin, inevitably, by measuring the Earth.

And one way of doing that is to climb a mountain of known height H and estimate the distance D to the horizon (Figure 12). As the line of sight PQ will be tangent to the Earth, it will be at right angles to the radius of the Earth, OQ, so OQP will be a right-angled triangle.

Applying Pythagoras' theorem, we have

$$(R+H)^2 = R^2 + D^2,$$

where R is the radius of the Earth. After rewriting the left-

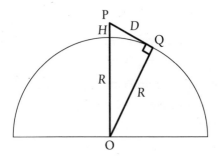

12. Measuring the Earth.

hand side as $R^2 + 2RH + H^2$ and cancelling the R^2 terms we have $2RH + H^2 = D^2$.

In practice, H will be tiny compared to the radius of the Earth R, so that H^2 will be tiny compared to $2RH$. Thus, $2RH$ is approximately equal to D^2, and so

$$R \approx \frac{D^2}{2H}.$$

In about 1019, the scholar Al-Biruni used broadly similar ideas to estimate the radius R of the Earth, obtaining a result which differed from the currently accepted value by less than 1%. This was a quite extraordinary achievement for the time.

Equations and curves

I should like to end this chapter by pointing out one particularly powerful way in which geometry and algebra come together.

Today, if we have a relationship between two numbers—$y = x^2$, for example—we think nothing of using x and y as *coordinates* to plot a graph, as in Figure 13. Our equation is then represented by a curve. And, conversely, if some problem in geometry involves a certain curve, we can try and represent it by an equation.

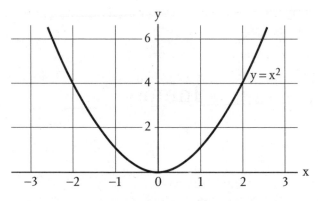

13. Coordinate geometry.

But in Newton's time this was a very new idea indeed, largely due to two French mathematicians, Pierre de Fermat (1601–65) and René Descartes (1596–1650).

And while it takes us very close to calculus itself, we need, first, just one more key idea…

3

Infinity

Infinity enters our story very early, around the time of Archimedes, in about 220 BC.

To be more precise, what really matters is the idea of gradually *approaching* infinity, and I would like to offer two examples.

The area of a circle

The two formulae in Figure 14 are among the best known in the whole of mathematics. But why are they true?

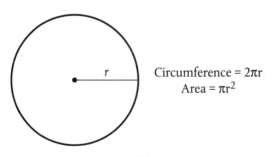

Circumference = $2\pi r$
Area = πr^2

14. Circle formulae.

Well, for the purposes of this book I would like to define π as

$$\pi = \frac{\text{circumference of circle}}{\text{diameter}},$$

because that ratio is the same for all circles.

So, if the radius is r, the diameter is $2r$, and the first result follows straight from the definition; it is, more or less, simply a re-statement of what we actually mean by the number π.

But the other formula, area $= \pi r^2$, is quite a different matter.

So, to see why it is true, let us follow Archimedes—rather loosely, in the first instance—by inscribing within the circle a regular polygon with N equal sides (Figure 15).

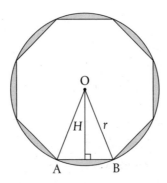

15. Approximating a circle.

Now, the polygon will consist of N triangles such as OAB, where O is the centre of the circle, and the area of each such triangle will be $\frac{1}{2}$ its 'base' AB times its 'height' H. The total area of the polygon will therefore be N times this, i.e. $\frac{1}{2} \times (AB) \times H \times N$.

But $(AB) \times N$ is the length of the perimeter of the polygon, so

$$\text{area of polygon} = \frac{1}{2} \times (\text{perimeter of polygon}) \times H.$$

The idea now is to get at the area of the circle itself by considering what happens *as N gets larger and larger*, so that the polygon has more and more sides (Figure 16).

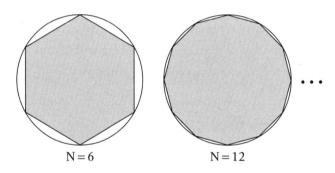

$N = 6$ $N = 12$

16. Closer and closer...

Plainly, as N increases, the perimeter gets closer and closer to the circumference of the circle, which is $2\pi r$.

And H gets closer and closer to r.

So the area of the polygon gets closer and closer to

$$\frac{1}{2} \times 2\pi r \times r,$$

which is πr^2.

The idea of a limit

I should admit at once that all this talk of 'getting closer and closer to' is, at best, a little vague.

More precisely, we may view the area of the circle itself as the *limit* of the polygon's area as $N \to \infty$, i.e. as N *tends* to infinity.

And, broadly speaking, what we mean by this is that we can make the difference between the two areas as small as we like by taking N large enough.

This idea of limit is central to the whole of calculus, but it is a subtle idea, and one which will gradually evolve and sharpen, I hope, during the course of this book.

Matters are not helped by the fact that the very word 'limit' is being used in a rather different way from that in which it is used in everyday life.

So, for the time being—and speaking very loosely indeed—a limit in mathematics is something that we can approach *as close as we like*, provided that we *try hard enough*.

An infinite series

Another way in which infinity enters our story is through the idea of *infinite series*, such as

$$\frac{1}{4} + \frac{1}{4^2} + \frac{1}{4^3} + \cdots = \frac{1}{3}.$$

Now, at first sight, this is quite remarkable. For, as the dots suggest, the series of positive terms on the left-hand side continues forever in the way indicated—yet the sum is finite, and just $\frac{1}{3}$.

For the time being, I simply offer a 'proof by picture' of this result, in which we take a square of side 1 and divide it up into a sequence of smaller and smaller squares (Figure 17).

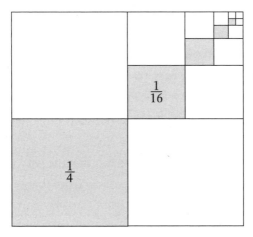

17. A 'proof by picture'.

The total shaded area then represents the sum of our infinite series, and it is evident, I think, that this area represents $\frac{1}{3}$ of the whole, because there is, essentially, an exact copy of it on either side.

Here again, however, there are subtleties, and the proof in Figure 17 is a little casual.

A better description of what the result

$$\frac{1}{4}+\frac{1}{4^{2}}+\frac{1}{4^{3}}+\cdots=\frac{1}{3}$$

really means is that we can make the running total on the left-hand side as close to $\frac{1}{3}$ as we like by taking enough terms.

In other words, $\frac{1}{3}$ represents the limit of that running total as the number of terms, N, tends to infinity.

The road to calculus

Armed with all we have seen so far, and some concept of *limit* in particular, we are now ready to embark properly on our journey.

And the road to calculus involves four main themes:

  (i) the steepness of a curve,
 (ii) the area enclosed by a curve,
(iii) infinite series, and
(iv) the problem of motion.

We will look at each of these, in turn, in Chapters 4–12, and I hope, of course, that I will succeed in explaining the key ideas as simply and clearly as possible.

But I am not claiming that calculus is ever easy. It isn't.

One reason I know this is a visit to my father some years ago, just a few weeks before he died.

He was not a mathematician, but he had kindly offered to comment on something I was writing at the time.

And we were sitting comfortably, looking out on the evening sun in his back garden, when he suddenly said:

'I'm afraid I don't agree with you that $\frac{1}{4}+\frac{1}{16}+\frac{1}{64}+\ldots$ is equal to $\frac{1}{3}$. I believe it is less than $\frac{1}{3}$, *by an infinitely small amount*.'

In reply, I said:

'I might be tempted to agree with you, *if* I knew what it means for a number to be infinitely small. But I don't.'

'Ah!' he said, most thoughtfully, and I immediately began to marshal my own thoughts in preparation for a counter-attack.

But, in the end, none came, and all he eventually said was:

'Let's have another glass of whiskey!'

4

How Steep is a Curve?

Calculus is all about the rates at which things change.

And, as we have seen already, this idea is related to the steepness of a curve.

So, how *do* we determine the steepness, or slope, of a curve at any particular point?

The slope of a straight line

In the case of a straight line, the answer is simple: we just take two points P and Q on the line, and calculate the increases in x and y as we move from P to Q (Figure 18). Then

$$\text{slope} = \frac{\text{increase in } y}{\text{increase in } x}.$$

The great merit of this definition is that it doesn't matter which two points of the line we choose—this *ratio* is always the same.

And—fairly evidently, I think—the larger the ratio, the steeper the line.

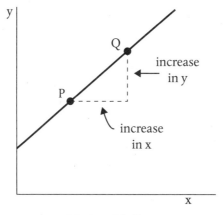

18. A straight line.

The slope of a curve

But if we try to apply this same idea to determine the slope of a curve at some point *P*, we hit a problem, for the ratio

$$\frac{\text{increase in } y}{\text{increase in } x}$$

will typically depend on where we choose our second point, *Q*.

So, where should we choose *Q*?

As we are trying to determine the steepness of the curve *at the point P*, rather than somewhere else, we should presumably choose *Q* so that it is close to *P*.

But how close, exactly?

And, after a little more thought still, the natural answer would seem to be: *the closer the better* (Figure 19).

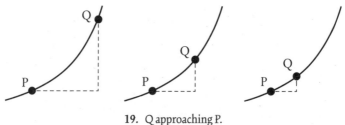

19. Q approaching P.

In this way, then, we are led to define the slope of the curve at P as the *limit* of the ratio as Q *tends* to P:

$$\text{slope of curve at } P = \lim_{Q \to P} \frac{\text{increase in } y}{\text{increase in } x}.$$

An example

The simplest way to see this idea in action is, I think, with the curve $y = x^2$ (Figure 20).

So, if the x-coordinates of P and Q are x and $x + h$, say, then the corresponding y-coordinates will be x^2 and $(x + h)^2$. And as we move from P to Q there is therefore an increase in y of amount $2xh + h^2$ (by Chapter 2).

It follows, then, that

$$\frac{\text{increase in } y}{\text{increase in } x} = \frac{2xh + h^2}{h},$$

and on cancelling the factors of h we have

$$\frac{\text{increase in } y}{\text{increase in } x} = 2x + h.$$

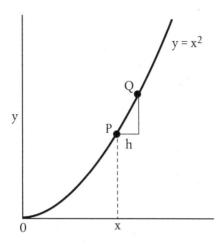

20. Finding the slope of a curve.

Finally, then, we fix the point P—and hence the coordinate x—and take the limit Q → P, that is h → 0, giving

$$\text{slope of curve } y = x^2 = 2x.$$

So the slope increases with x, and this makes sense, of course, because the curve $y = x^2$ evidently 'bends upward', and therefore gets steeper as x increases.

The whole procedure which we have just described is fundamental to calculus, for two reasons.

From a purely geometrical point of view, it lets us construct the *tangent* to the curve at any point, because the slope of the curve at that point will be the slope of the tangent (Figure 21).

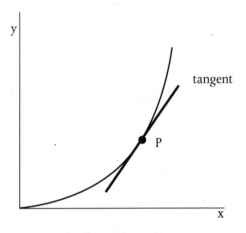

21. The tangent to a curve.

From a dynamical point of view, on the other hand, it lets us calculate the rate at which y increases with x, because that is precisely the slope of the curve.

And this whole procedure of obtaining the slope of a curve, from its equation, is called *differentiation*.

5

Differentiation

The whole idea of differentiation is so central to calculus that there is a special notation for it.

First, the Greek letter δ, i.e. 'delta', denotes not a number but the phrase 'increase in …'. So, for example, if x were to increase from 1 to 1.01, then δx would be 0.01.

In this way, then, δx and δy denote the small increases in x and y that occur as we move along some curve from the point P to a nearby point Q (Figure 22).

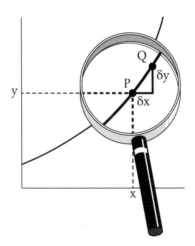

22. Small increases in x and y.

Now, as we have seen, the whole process of differentiation involves finding the *limit* of $\delta y / \delta x$ as $\delta x \to 0$, i.e. *as* Q *moves closer and closer to* P.

And we now denote this limit by the special symbol dy/dx, as in Figure 23.

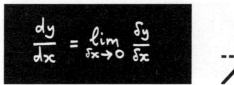

23. Definition of dy/dx.

This entity—pronounced '*d y d x*'—is called the *derivative* of *y* with respect to *x*, and represents both the slope of the curve and the rate at which *y* is increasing with *x*.

The distinctive notation, due to Leibniz, has proved superbly successful over the years, but there are some subtleties.

There seems no doubt that—in his earlier years, at least—Leibniz viewed dy/dx as the ratio of two numbers, dy and dx, both of which were 'infinitely small'.

We will *not* attempt to do this, but, instead, will consistently view it as the limit of the genuine ratio $\delta y / \delta x$ as $\delta x \to 0$.

Indeed, if we 'deconstruct' the symbol dy/dx at all, it will tend to be in the following way:

$$\frac{d}{dx}(y),$$

where we are viewing d/dx itself as a symbol, meaning 'differentiate with respect to *x*'.

Examples

Now, we saw in Chapter 4 how to actually do all this, and we already know from there how to differentiate $y = x^2$ (Figure 24).

$$\frac{d}{dx}(x^2) = 2x$$

24. The derivative of x^2.

I now offer a second example, if only to show the dy/dx notation in action.

Suppose, then, that $y = 1/x$.

Notably, as x increases, y decreases in this case (see Figure 25), so we might reasonably expect a negative slope, and therefore a negative value of dy/dx.

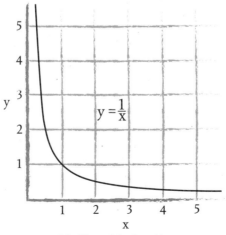

25. The graph of $y = 1/x$.

In any event, our first task is to calculate the quantity δy. And when the x-coordinate changes from x to $x + \delta x$, y will change from $1/x$ to $1/(x + \delta x)$, so

$$\delta y = \frac{1}{x + \delta x} - \frac{1}{x}.$$

By the usual rules of algebra, this may be rewritten as

$$\delta y = \frac{-\delta x}{(x + \delta x)x},$$

so that

$$\frac{\delta y}{\delta x} = -\frac{1}{(x + \delta x)x}.$$

On finally letting $\delta x \to 0$, we find that $dy/dx = -1/x^2$.

We have therefore shown that

$$\frac{d}{dx}\left(\frac{1}{x}\right) = -\frac{1}{x^2},$$

and the derivative is indeed negative in this case, as we anticipated.

In the same general way, we can gradually build up the collection of results shown in Figure 26.

And if, by any chance, you feel that there might be a pattern developing here, with the derivative of x^4 being $4x^3$, and so on, you are in fact quite right; the derivative of x^n is given by Figure 27 for any positive whole number n, and we will explain why later, in Chapter 13.

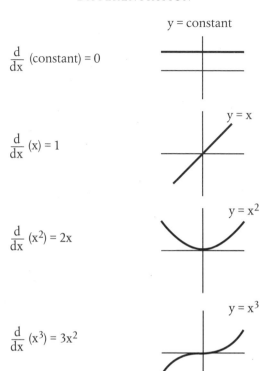

$$\frac{d}{dx} (\text{constant}) = 0$$

$y = \text{constant}$

$$\frac{d}{dx} (x) = 1$$

$y = x$

$$\frac{d}{dx} (x^2) = 2x$$

$y = x^2$

$$\frac{d}{dx} (x^3) = 3x^2$$

$y = x^3$

26. Some elementary derivatives.

$$\frac{d}{dx} (x^n) = nx^{n-1}$$

27. Differentiating x^n.

Functions

In all the examples in the previous section there is just one, unique value of y corresponding to each given value of x.

Whenever this is the case, we say that y is a *function* of x.

Thus, $y = x^2$ defines y as a function of x, but it does not define x as a function of y, because any given (positive) value of y leads to two possible values for x, one positive and one negative.

Two general rules

In addition to the specific results we've been discussing, there are two general rules which are very helpful:

$$1. \ \frac{d}{dx}(u + v) = \frac{d}{dx}(u) + \frac{d}{dx}(v).$$

2. If c is a *constant*, then
$$\frac{d}{dx}(cy) = c\frac{d}{dx}(y).$$

Here, u, v, and y can be any functions of x which can be differentiated.

In Chapter 6, for instance, we will find ourselves wanting to differentiate $4x - 2x^2$. Rule 1 says that we can differentiate $4x$ and $-2x^2$ separately, and simply add the results. And rule 2 says that the derivative of $4x$ is just 4 times the derivative of x, i.e. $4 \times 1 = 4$. In a similar way, the derivative of $-2x^2$ is $-2 \times 2x = -4x$.

While a little technique of this kind will be needed in the pages which follow, the more pressing question, surely, is: what is all this differentiation *for*?

6

Greatest and Least

One major use of calculus is in problems of *optimization*, where we have some quantity, and want to find its maximum or minimum value.

Down on the farm...

Imagine, for instance, that you are a farmer, and you want to create a rectangular field next to a river (Figure 28). Suppose, too, that you have a fixed amount of fencing—say 4 km—for the other three sides.

How should you arrange things so that the area A of the field is *as large as possible*?

Should you, for example, choose the rectangle so that it is a square?

Now, I should confess at once that I have never actually met a farmer who wanted to do anything of the kind, but this little problem does illustrate well one particular aspect of calculus in action.

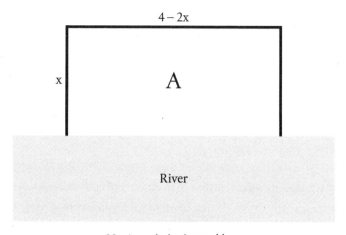

28. A maximization problem.

To see this, let x denote the width of the field, so that the side parallel to the river must be of length $4 - 2x$.

The area of the field will therefore be $x(4 - 2x)$, so

$$A = 4x - 2x^2,$$

and our problem is to choose x so that A is a maximum.

And the key step is to differentiate with respect to x, which gives

$$\frac{dA}{dx} = 4 - 4x.$$

Now, plainly, if $x < 1$ then dA/dx is positive and A increases with x, but if $x > 1$ then dA/dx is negative and A decreases with x.

Not only does this help us sketch the graph of A against x (Figure 29), but it tells us, of course, that the maximum value of A must occur when

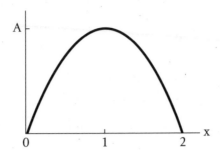

29. How *A* depends on *x*.

$$\frac{dA}{dx} = 0,$$

i.e. when $x = 1$, because this is where *A* stops increasing with *x* and starts decreasing.

And when $x = 1$, the side parallel to the river, namely $4 - 2x$, is equal to 2. So we maximize the area by choosing a rectangle with an aspect ratio of 2:1 (Figure 30).

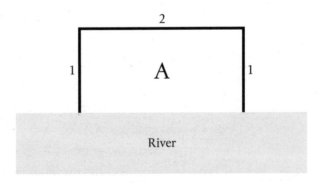

30. The solution to the problem.

But, more generally...

The idea of tackling optimization problems by differentiating is a powerful one, due essentially to Fermat in about 1630, but there are subtleties.

In some problems, for instance, setting $dy/dx = 0$ will deliver the *minimum* value of y (Figure 31a).

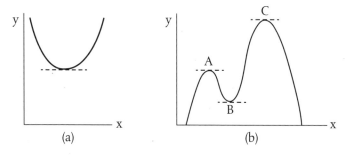

31. (a, b). Some optimization problems.

More generally still, it might be that the graph of y against x looks something like Figure 31b. Setting $dy/dx = 0$ will then yield three values of x, corresponding to the points A, B, and C, and further work will be required to show that C gives the maximum value of y and that none of them give its minimum value, over the range of x shown in the figure.

So setting $dy/dx = 0$ is only ever part of the story.

What's the best view of Nelson's Column?

I should like to end this chapter with one of my favourite optimization problems, even though the details require rather more technique than we have developed so far.

Imagine, then, that you are in Trafalgar Square, London looking up at Nelson's column.

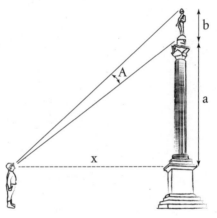

32. What's the best view?

Clearly, if you stand too far away your viewing angle A will be very small, but it will also be small if you stand too close, because you will then be viewing Nelson very obliquely.

So, at what distance x should you stand to maximize A?

Calculus—eventually—gives the answer:

$$x = \sqrt{a(a+b)},$$

where b is Nelson's height and a the distance of his feet above your eyeline.

And because, in practice, b is small compared to a, this implies that you should look up at an angle of about 45°.

But watch out for the traffic!

7

Playing with Infinity

Earlier, we proved that the area of a circle is πr^2, by using an N-sided polygon and letting $N \to \infty$.

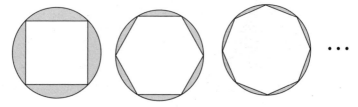

33. Approximations to a circle.

But while we attributed this whole idea to Archimedes, it is not exactly what Archimedes does.

He begins, instead, by assuming that the area is *greater* than πr^2. He then introduces an inscribed polygon, as we did in Chapter 3, and shows that a contradiction arises for some sufficiently large, but finite, value of N.

He then tries the assumption that the area is less than πr^2, draws a polygon touching the *outside* of the circle, and shows that another contradiction arises for some sufficiently large N.

The only possibility left, then, is that the area of the circle is exactly πr^2.

And this whole line of argument is called *reductio ad absurdum*, or 'proof by contradiction'.

The precise way in which the contradictions arise need not concern us here. The real point is that at no stage in the argument is N just allowed to tend to infinity—let alone be infinite; the number of polygon sides, N, is always finite.

In a broadly similar way, Archimedes proves the results for a sphere shown in Figure 34, and the result for a cone dates from even earlier work by Eudoxus (*c.*360 BC).

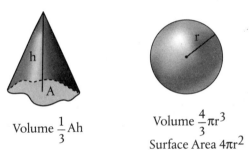

Volume $\frac{1}{3}Ah$

Volume $\frac{4}{3}\pi r^3$

Surface Area $4\pi r^2$

34. Cone and sphere formulae.

But, again, at no stage is anything allowed to just 'tend to infinity'.

In their final, polished proofs, at least, the ancient Greeks avoided infinity like the plague.

Mathematicians living dangerously

By 1615 things had changed, and the German astronomer Johannes Kepler was apparently quite happy to regard a sphere as an infinite number of infinitely thin cones extending from its centre (Figure 35).

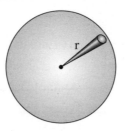

35. Kepler's approach to the volume of a sphere.

In this way, he reasoned, it is easy to obtain the volume of a sphere from the formula for its surface area.

After all, the volume of each cone is $\frac{1}{3}r$ times its base area, and the base areas of all the infinitely thin cones add up to the surface area of the sphere, $4\pi r^2$.

So the volume of the sphere must be

$$\frac{1}{3}r \times 4\pi r^2 = \frac{4}{3}\pi r^3,$$

mustn't it?

A little later, Bonaventura Cavalieri, who had been a student of Galileo, came up with an ingenious new approach to areas and volumes.

In Figure 36, for example, the two geometrical shapes have (a) the same height and (b) the same width, or horizontal extent, *at every level*.

According to Cavalieri, then, the two shapes must have the same area.

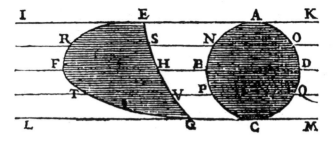

36. From Cavalieri's *Exercitationes Geometricae Sex* (1647).

(To take a loose analogy: we do not change the volume of a deck of playing cards simply by displacing some of them.)

Cavalieri's principle makes it possible to calculate the area (or volume) of some awkwardly shaped object by reference to a much simpler one.

He appears to be regarding an area as composed of infinitely many lines, but what Cavalieri really tries to do, in effect, is sidestep the matter of infinity altogether.

Even later still, John Wallis, Savilian professor of mathematics at Oxford, threw caution completely to the wind and embraced infinity with sufficient confidence that he even invented a symbol for it: ∞.

Wallis was a brilliant mathematician, as indicated by the following extraordinary infinite product for π,

$$\frac{\pi}{2} = \frac{2}{1} \times \frac{2}{3} \times \frac{4}{3} \times \frac{4}{5} \times \frac{6}{5} \times \frac{6}{7} \times \frac{8}{7} \times \frac{8}{9} \cdots$$

which he discovered in 1655. But some of the things he did would now be viewed as downright dangerous.

4   *De Sectionibus Conicis.*   **PROP. 1.**

PARS PRIMA.

PROP. I.

*De Figuris planis juxta Indivisibilium
methodum confiderandis.*

 Uppono in limine (juxtâ Bonaventuræ
Cavallerii *Geometriam Indivisbilium*)
Planum quodlibet quafi ex infinitis lineis
parallelis conflari: Vel potiùs (quod e-
go mallem) ex infinitis Prallelogram-
mis æquè altis; quorum quidem fingulo-
rum altitudo fit totius altitudinis $\frac{1}{\infty}$, five aliquota pars
infinite parva; (efto enim ∞ nota numeri infini-
ti;) adeóq; omnium fimul altitudo æqualis altitudi-
ni figuræ.

37. First appearance of the infinity '∞' sign, in John Wallis's
De Sectionibus Conicis (1656).

In Figure 37, for instance, Wallis considers a parallelogram
whose height is 'infinitely little', and writes that height as 1/∞.
Elsewhere, he even writes

$$\frac{1}{\infty} \times \infty = 1.$$

Today, we view this as complete nonsense, and do not
regard ∞ as a *number* at all.

Even at the time, the philosopher Thomas Hobbes, who was a great admirer of Euclidean geometry, poured scorn on Wallis's whole approach, writing:

> I verily believe...that since the beginning of the world there has not been...so much absurdity written in geometry.

A safer approach

It is safer, surely, to approximate a curved region by a *finite* number of simple pieces, and then see what happens as that number gets bigger and the pieces themselves get smaller.

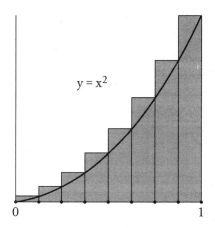

38. Approximating a curved region by rectangles.

Suppose, for example, that we want to find the area under the curve $y = x^2$, between $x = 0$ and $x = 1$. We can approximate the region by N rectangles, each of width $1/N$, as in Figure 38.

Then, with the help of the formula

$$1^2 + 2^2 + .. + N^2 = \frac{1}{6}N(N+1)(2N+1),$$

which has been known since the time of Archimedes, we can show that the shaded area in Figure 38 is

$$\frac{1}{6}\left(1+\frac{1}{N}\right)\left(2+\frac{1}{N}\right).$$

On finally letting $N \to \infty$, so that the rectangles get thinner and more numerous, and approximate the curved region ever more closely, we obtain $\frac{1}{3}$ for the area under the curve itself.

And in the 1630s Fermat and others used methods of this general kind to calculate many different areas with curved boundaries.

Yet there is, in fact, another way…

8

Area and Volume

His name is Mr Newton; a fellow of our College, & very young...but of an extraordinary genius and proficiency in these things.

Isaac Barrow, of Trinity College Cambridge,
in a letter of 1669

Suppose that we want to find the area A under some curve.

Plainly, if we change x, then A will also change, and Newton showed that it does so, in fact, in the way shown in Figure 39.

39. The fundamental theorem of calculus.

And this result, called the *fundamental theorem of calculus*, is really quite extraordinary. After all, we have seen that—

geometrically at least—differentiation is all about finding the steepness of a curve.

Now we find that *undoing* differentiation is a way of finding area.

I say 'undoing' because, in practice, we will usually know y as a function of x in the equation in Figure 39, and will be trying to find A.

And this process of undoing, or reversing, differentiation is called *integration*.

Here's a simple example.

The area under y = x², revisited

In this case, evidently,

$$\frac{dA}{dx} = x^2,$$

And so, to determine A, we find ourselves asking what function of x, when differentiated, gives x^2.

Well, a glance back at Chapter 5 reminds us that the derivative of x^3 is $3x^2$—which is getting close—and the second 'general rule' then tells us that the derivative of $\frac{1}{3}x^3$ will be x^2.

At this point, a little care is needed, because $\frac{1}{3}x^3$ is not the only function of x with derivative x^2. The derivative of a constant is zero, so we could add any constant c and still have the derivative dA/dx equal to x^2:

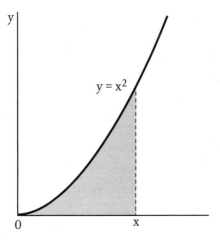

40. The area under $y = x^2$.

$$A = \frac{1}{3}x^3 + c.$$

As it happens, if we are measuring the area A under the curve from $x = 0$, as in Figure 40, we require that $A = 0$ when $x = 0$. In our case, then, c is in fact zero, and $A = \frac{1}{3}x^3$ emerges as the final answer.

In particular, on putting $x = 1$ we find that the area under the curve $y = x^2$ between $x = 0$ and $x = 1$ is $\frac{1}{3}$, in agreement with the conclusion of Chapter 7.

Proof of dA/dx = y

To see why all this works, turn back, if you will, to Figure 39 and imagine x increasing very slightly to $x + \delta x$. Then A will increase very slightly also, and the additional area will be a tall, thin strip of width δx, as shown in Figure 41a.

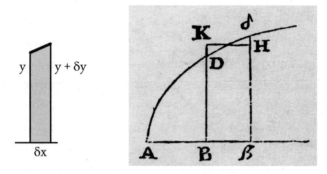

41. (a) A small increase in area. (b) From Newton's *De Analysi* (1669, published 1711).

Suddenly, then, it is rather easy to see, in rough terms, why $dA/dx = y$, because this additional area is, very nearly, a long, thin rectangle of width δx and height y, so that, very nearly, $\delta A = y \, \delta x$.

But we can sharpen our argument by borrowing an idea from an early Newton manuscript of 1669, commonly known as *De Analysi* (see Figure 41b).

Not surprisingly, the notation there is quite different: A, D, and δ denote points on the curve, and $B\beta$ corresponds to our δx. But Newton observes that the additional area will be *exactly* that of some rectangle $B\beta HK$ with width δx and height—in our terms—*somewhere between y and $y + \delta y$*.

So, in our terms, $\delta A/\delta x$ is sandwiched firmly between y and $y + \delta y$. And in this way, then, if we finally let $\delta x \to 0$ (so that $\delta y \to 0$ also) we obtain the result: $dA/dx = y$.

Torricelli's trumpet

The same general line of reasoning can be used to find volumes, and the cone and sphere formulae in Chapter 7 can certainly be established by calculus, i.e. integration methods.

But I would like to consider instead a rather more exotic example.

In 1643, Evangelista Torricelli, another mathematician who had studied with Galileo, caused quite a sensation by discovering a three-dimensional object that had *infinite extent but finite volume*.

Even 30 years later, when Thomas Hobbes heard of this result, he wrote:

> to understand this for sense, it is not required that a man should be a geometrician or a logician, but that he should be mad.

But was Torricelli right?

To find out, we can use some calculus.

His example was a trumpet-shaped object, which we can obtain by rotating the curve $y = 1/x$ about the x-axis, all the way from $x = 1$ to infinity (Figure 42).

Now, the volume V of the shaded region (measured from the end of the trumpet, at $x = 1$) will plainly depend on x, and if we increase x by a small amount to $x + \delta x$, the additional volume δV will be, essentially, a thin circular disc of radius y and thickness δx. The area of this disc will be πy^2, so we will have, very nearly, $\delta V = \pi y^2 \, \delta x$.

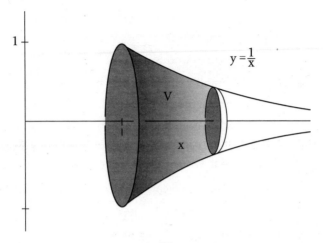

42. Torricelli's trumpet.

In this way, then, we conclude that V must depend on x in such a way that

$$\frac{dV}{dx} = \pi y^2,$$

and this is, essentially, a three-dimensional equivalent of the equation $dA/dx = y$ at the beginning of this chapter.

Now, in our particular case, $y = 1/x$, so

$$\frac{dV}{dx} = \frac{\pi}{x^2},$$

and our recent experience (and another glance back at Chapter 5) allows us to integrate this fairly immediately:

$$V = -\frac{\pi}{x} + c,$$

where c is a constant.

And, this time, the constant isn't zero; we know that V must be zero at the end of the trumpet, $x = 1$, because it is measured from there. So $c = \pi$, and our final answer is

$$V = \pi\left(1 - \frac{1}{x}\right).$$

And when we take the limit $x \to \infty$, corresponding to the whole trumpet, $V \to \pi$, which is most certainly finite.

So Torricelli was right.

Would you believe it?

Calculus provides many other surprises concerning area and volume, though they are not always of great practical value.

Spherical bread

If the slices of a spherical loaf of bread are of equal thickness, which piece has the most crust?

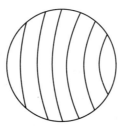

43. Spherical bread.

The answer, surprisingly, is that they all have the same amount of crust (i.e. surface area), and this result was known to Archimedes.

The pizza theorem

Take any internal point P of a circle, and make two cuts at right angles to each other. Then make a further two cuts bisecting the angles made by the first two.

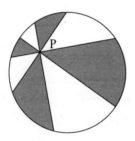

44. Sharing pizza.

The four shaded pieces will then have the same total area as the four unshaded pieces, making this an exotic way of sharing a pizza equally.

To the Earth's core...

A cylindrical hole, of depth L, is drilled through a sphere, passing straight through its centre.

What volume of material is left?

Answer: $\frac{1}{6} \pi L^3$, *regardless of the size of the sphere.*

So if you drill a hole of depth 6 cm through a sphere the size of an apple you will have 36π cubic cm left over.

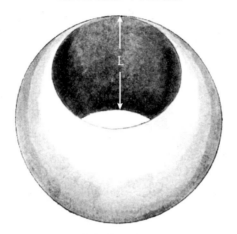

45. A hole through a sphere.

And if you bore a hole of depth 6 cm through a sphere the size of the *Earth* you will again have 36π cubic cm left.

At first, perhaps, this seems incredible, until we realize that with a hole of *depth* 6 cm there won't, indeed, be much of the 'Earth' left—just a very thin ring around the equator.

9

Infinite Series

We have already seen that an infinite series can have a finite sum:

$$\frac{1}{4} + \frac{1}{4^2} + \frac{1}{4^3} + \ldots = \frac{1}{3}.$$

But in order to apply this idea to calculus we need to broaden our scope a little, and consider series in which the individual terms are functions of x.

The simplest of these is the so-called *geometric series*:

$$1 + x + x^2 + x^3 + \ldots = \frac{1}{1-x} \qquad \text{for} -1 < x < 1$$

And there is a remarkably easy way of proving this particular result.

We start by writing down the sum s_n of the first n terms, and then multiply by x:

$$s_n = 1 + x + x^2 + \cdots + x^{n-1}$$
$$xs_n = x + x^2 + \cdots\cdots + x^n.$$

On subtracting, there is a spectacular amount of cancellation on the right-hand side, and we are left with

$$(1-x)s_n = 1 - x^n.$$

Finally, we take the limit as $n \to \infty$. Provided $-1 < x < 1$, we then find that $x^n \to 0$, so that $s_n \to 1/(1-x)$, which is what we were trying to show.

Closer and closer…

Setting $x = 1/4$, for instance, leads to the result at the beginning of this chapter.

Setting $x = -1/2$, on the other hand, produces a series with alternating signs:

$$1 - \frac{1}{2} + \frac{1}{4} - \frac{1}{8} + \cdots = \frac{2}{3}.$$

The running total, s_n, therefore oscillates (see Figure 46), but convergence is fast, so that s_n gets quite close to the limit 2/3 after just six or seven terms.

But these are just special cases. We have shown that the series

$$1 + x + x^2 + x^3 + \cdots$$

converges to the sum $1/(1-x)$ for *any* value of x in the range $-1 < x < 1$.

And there is, I think, a slight danger at this point that the condition on x may appear rather 'obvious'. It does, after all,

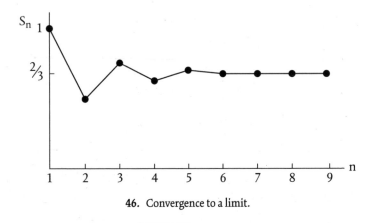

46. Convergence to a limit.

ensure that the individual terms get smaller (rather than bigger) in magnitude as the series goes on.

In fact, however, convergence can be a very subtle matter, and it turns out that, more generally, getting smaller *isn't enough*.

A divergent series

Consider, for instance,

$$1+\frac{1}{2}+\frac{1}{3}+\frac{1}{4}+\frac{1}{5}+\cdots.$$

Here, again, the individual terms get steadily smaller, yet in this case the series has no finite sum, for we can make the running total as large as we like by taking enough terms.

This was proved as long ago as 1350 by the French scholar Nicole Oresme, and the proof itself is stunning in its simplicity. He just groups the terms, after the first, in the following way:

$$\frac{1}{2}$$

$$\frac{1}{3} + \frac{1}{4}$$,

$$\frac{1}{5} + \frac{1}{6} + \frac{1}{7} + \frac{1}{8}$$

$\vdots$

so that each new group has twice as many terms as the previous one.

Oresme then observes that $\frac{1}{3} + \frac{1}{4}$ is greater than $\frac{1}{4} + \frac{1}{4} = \frac{1}{2}$, that the next group is greater than $\frac{1}{8} + \frac{1}{8} + \frac{1}{8} + \frac{1}{8} = \frac{1}{2}$, that the one after that is greater than $8 \times \frac{1}{16} = \frac{1}{2}$, and so on, for ever.

And as $\frac{1}{2} + \frac{1}{2} + \frac{1}{2} + \ldots$ doesn't converge to a finite sum, it follows that the series in question can't either.

This, then, is something of a cautionary tale, with important consequences that we will see later.

But I would like to end this chapter on a quite different note. For it turns out that this particular result has a practical application—albeit a rather exotic one.

Extreme box-stacking

Imagine stacking some boxes, one on top of another, so that the column leans, somewhat perilously, over the edge of a table.

If each box has length 1 unit, how big can the overhang be before the whole column topples over, under gravity?

With just one box, evidently, the maximum overhang is $\frac{1}{2}$, but with four boxes this climbs to

$$\frac{1}{2}\left(1+\frac{1}{2}+\frac{1}{3}+\frac{1}{4}\right),$$

which is a little greater than 1, so that no part of the top box is directly over the table (Figure 47).

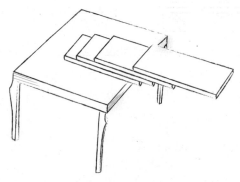

47. 4-box overhang.

And if we want an overhang of more than two box lengths, then we can just achieve that with 31 boxes, because the maximum overhang in this case turns out to be

$$\frac{1}{2}\left(1+\frac{1}{2}+\cdots+\frac{1}{31}\right)=2.0136$$

(see Figure 48).

And it goes on like this, with the maximum possible overhang with n boxes being

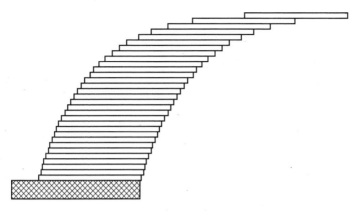

48. 31-box overhang.

$$\frac{1}{2}\left(1+\frac{1}{2}+\cdots+\frac{1}{n}\right).$$

Somewhat surprisingly, then, it turns out that we can make the overhang *as big as we like*—if only we have enough boxes—because the infinite series

$$1+\frac{1}{2}+\frac{1}{3}+\frac{1}{4}+\cdots$$

diverges, with no finite sum.

I have to admit, however, that I had never appreciated just how slowly it diverges until I once found myself in a maths show, with a lot of pizza boxes, in a major city-centre theatre.

Before the show began I calculated, just out of interest, how high the column of pizza boxes would have to be—on the above model—to overhang right across the stage.

The answer turned out to be 5.8 light years.

10

'Too Much Delight'

Integration, or undoing differentiation, is often quite challenging, and can require considerable ingenuity.

But infinite series can help, and, to see how, let us follow in Newton's footsteps for a moment, and try to determine the area under the curve in Figure 49, between 0 and x.

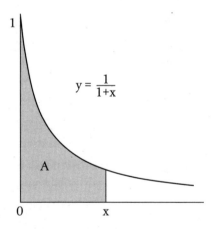

49. Area under a hyperbola.

We can, of course, start by writing

$$\frac{dA}{dx} = \frac{1}{1+x}.$$

But what function of x do we know that, when differentiated, gives $1/(1 + x)$? Our limited repertoire so far, in Chapter 5, certainly doesn't contain the answer—or give much of a clue.

Yet there is a way forward—if we rewrite the function $1/(1 + x)$ *as an infinite series*:

$$\frac{1}{1+x} = 1 - x + x^2 - x^3 + \cdots$$
$$\text{for} -1 < x < 1.$$

While it may look a little different, this is in fact the same mathematical result as that in Chapter 9. (Setting $x = 1/2$ here, for instance, gives the same outcome as setting $x = -1/2$ there.)

Now, integrating simple *powers* of x is relatively easy, because we know from Chapter 5 the following:

y	dy/dx
x	1
x^2	$2x$
x^3	$3x^2$
x^4	$4x^3$
$\vdots$	

So integrating x gives $x^2/2$ (plus a constant), integrating x^2 gives $x^3/3$, and so on.

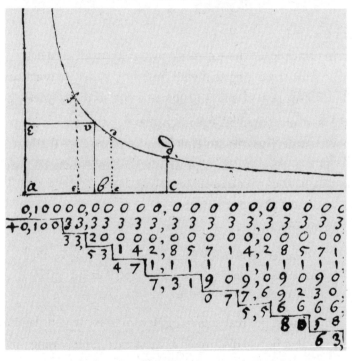

50. Two details from an early (*c.* 1665) Newton manuscript, concerning
the area under the hyperbola $y = a^2/(a+x)$. Only a small part
of his enormous calculation (with $a = 1$) is shown.

In this way, then, we can take the new form of our equation

$$\frac{dA}{dx} = 1 - x + x^2 - x^3 + \cdots$$

and integrate it term by term. And on applying the condition that $A = 0$ when $x = 0$ we obtain

$$A = x - \frac{x^2}{2} + \frac{x^3}{3} - \frac{x^4}{4} + \cdots$$

for $0 < x < 1$.

In principle, then, we can calculate A to any desired degree of accuracy by taking enough terms of the series. In practice, this will work best when x is quite small, so that successive terms get smaller quite quickly.

In Figure 50 we see some details from a very early manuscript by Isaac Newton in which he does, essentially, just this. In fact, he tries to calculate the area between $x = 0$ and $x = 0.1$ to an almost absurd accuracy.

In his own words:

> in summer 1665 being forced from Cambridge by the Plague 1 computed the area of the Hyperbola at Boothby in Lincolnshire to two and fifty figures

In fairness, the real source of Newton's excitement lay in the fact that he had discovered a *general* method for doing this kind of thing, as we will see later in the book.

Even so, several years later, he himself wrote:

> I am ashamed to tell to how many places I carried these computations, having no other business at the time: for then I took really too much delight in these inventions…

11

Dynamics

We now turn to the fourth and final strand in the early development of calculus, namely *dynamics*.

And it is only natural to revisit, first, the falling apple of Chapter 1. We saw there that the distance fallen, s, is proportional to t^2, and it is usually written in the manner shown in Figure 51.

$$s = \tfrac{1}{2}gt^2$$

51. The falling apple, revisited.

The constant g has a special significance, and we can use calculus, quite easily, to see what this is.

Note first that the downward velocity of the apple, v, is simply the rate at which the distance s increases with time. So $v = ds/dt$, and as the derivative of t^2 is $2t$ we find that

$$v = gt.$$

So the downward velocity of the apple increases with time, as we observed at the start of the book.

Moreover, acceleration is simply the rate at which velocity changes with time, and this is

$$\frac{dv}{dt} = g.$$

So the constant g denotes the downward *acceleration due to gravity*, which is approximately 9.81 m s^{-2}.

Velocity and acceleration

Before going any further, I should perhaps emphasize the distinction, in both mathematics and science, between speed and velocity.

Speed is simply a positive number, but velocity is a *vector* quantity, and therefore has both magnitude *and direction*.

So the two motions represented schematically in Figure 52 have the same speed but different velocity, because the two motions are in different directions.

This distinction becomes even more important as soon as we start talking about acceleration.

52. Two different velocities.

When travelling in a car, for instance, we tend to think of acceleration as rate of change of speed, without regard to the direction of motion.

But this is, in fact, mathematically and scientifically inaccurate. Acceleration isn't rate of change of speed; it's rate of change of *velocity*.

So, even if an object is moving at constant speed, it will have a nonzero acceleration if the direction of motion is changing.

And acceleration itself, like velocity, is a vector quantity, with both magnitude and direction.

Force and acceleration

The reason that acceleration is so important in dynamics is that, for an object of constant mass:

$$\text{Force} = \text{Mass} \times \text{Acceleration}.$$

This fundamental law of dynamics is due, essentially, to Newton, though it was never actually stated by him in precisely these terms.

53. Defying gravity.

Figure 53, for instance, shows some people on the inside of a giant, rapidly rotating drum at a funfair in the 1950s. And the only reason they don't fall down is that a large friction force at the wall holds them up against gravity.

This friction force is itself a consequence of the way in which the wall exerts a large force *inward*, towards the rotation axis, on each mass *m* as it moves along its circular path.

And the reason for that inward force – however strange this might seem at first sight – is that each object (and person) in the picture is *continually accelerating towards the centre of the circle.*

Circular motion

When an object moves at constant speed v around a circle of radius r it has an acceleration v^2/r towards the centre.

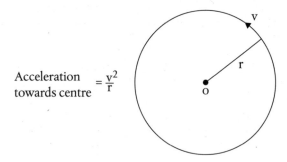

$$\text{Acceleration towards centre} = \frac{v^2}{r}$$

54. Acceleration in circular motion.

To demonstrate this, our approach will be calculus-like, in the sense that we will suppose the object to be at a certain point, and then see where it is a very short time later.

Suppose, then, that it is at the point P at a certain instant (Figure 55).

Now, if it were not accelerating, it would have to continue at the same old speed *in the same old direction*, i.e. along the tangent at P, so that it would be at the point R a time t later, having travelled a distance vt.

Let Q be the point where the straight line OR meets the circle.

Suppose now that the elapsed time t is very small, so that vt is much smaller than r. The points Q and R will then be very close to P.

In that case (and only then) the distances PQ and PR will be very nearly equal, so that at time t the object—which has been travelling at speed v *round the circle*—will be, very nearly, at the point Q.

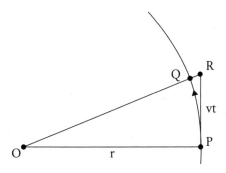

55. Proof that acceleration $= v^2/r$.

It will therefore have 'fallen' a distance QR towards O, and because QR is tiny compared to r the final formula in Chapter 2 applies, and (with a little rearrangement) tells us that QR $= (vt)^2/2r$.

But this can be rewritten in the form

$$QR = \frac{1}{2}\left(\frac{v^2}{r}\right)t^2,$$

and we see at once that this is precisely the $\frac{1}{2}gt^2$ formula for the falling apple, but with a different constant factor—v^2/r instead of g.

And this is why v^2/r represents the acceleration, towards the centre, in circular motion.

12

Newton and Planetary Motion

There goes the man that writt a book that neither he nor anybody else understands.

remark by a student at Cambridge, soon after the publication of Newton's *Principia* (1687)

The story of planetary motion is one of the greatest in the history of science, and the central ideas of calculus play a key part, albeit in a slightly hidden way.

Yet it all begins, really, in ancient Greece, with the geometry of an ellipse.

To draw an ellipse, mark out two *focal points* H and I, and run a loop of string around them. Then keep moving the point E—as in Figure 56—while keeping the string taut.

If the loop of string is very long the resulting ellipse will be almost circular, but if it barely stretches round the two focal points the ellipse will be very long and thin.

And just in case this all seems incredibly remote from the whole idea of planetary motion, it *was*—until...

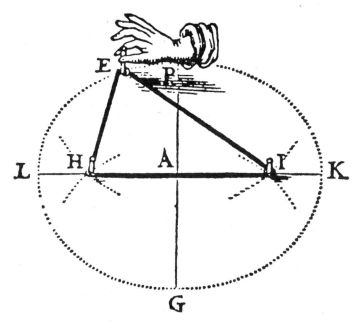

56. An ellipse, from van Schooten's *Exercitationum Mathematicorum* (1657).

Kepler's laws

In 1609, after an extraordinarily painstaking analysis of the astronomical observations of the planets, Johannes Kepler proposed the following:

1. The orbit of each planet is an ellipse, with the Sun at one focus.
2. A line drawn from the Sun to a planet sweeps out equal areas in equal times.

The first law, then, is about the shape of the orbit, and the second about the variation in speed as a planet goes round its

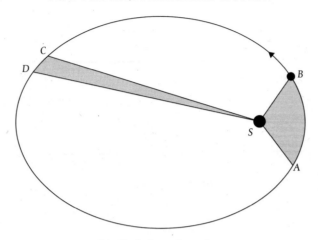

57. Kepler's equal-area law.

orbit, moving faster when close to the Sun and slower when further out, so that the areas swept out in a given time are the same.

Orbital data for the six planets known in Kepler's time

	$\bar{r}$ (units of $\bar{r}_{Earth}$)	T (years)
Mercury	0.387	0.241
Venus	0.723	0.615
Earth	1.000	1.000
Mars	1.524	1.881
Jupiter	5.203	11.862
Saturn	9.539	29.46

Somewhat later, in 1619, Kepler added a third law:

3. The orbital times T of the different planets increase with $\bar{r}$, the mean distance from the Sun, in such a way that

$$T^2 \text{ is proportional to } \bar{r}^3.$$

While we now see Kepler's laws as a landmark in the history of science, they were viewed with some scepticism in Newton's day. The second, area-sweeping law was regarded as particularly doubtful.

But the third law, T^2 proportional to $\bar{r}^3$, gained more general acceptance, and eventually helped point the way towards a gravitational theory of planetary motion.

An inverse-square law of gravitation?

In modern terms, the argument goes something like this.

The planetary orbits are only *slightly* elliptical, so if we approximate them by circles we can use Kepler's third law to work out how v, and hence v^2/r, depends on r.

Now, the orbital period—i.e. the time taken for each complete orbit—will be circumference divided by speed, $2\pi r/v$. So Kepler's third law implies that r^2/v^2 is proportional to r^3, so v^2 must be proportional to $1/r$.

That then implies that v^2/r, the acceleration towards O, must be proportional to $1/r^2$.

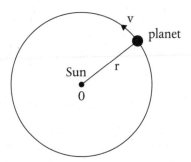

58. The circular approximation to planetary motion.

And, as acceleration is caused by a force, this suggests, at least, a force of attraction towards the Sun *which is proportional to* $1/r^2$.

Now, a calculation of this kind was certainly done by Newton, and possibly others, in the late 1660s, but the outcome will not have been so clear-cut, for two reasons.

First, the connection between force and (what we now call) acceleration had not been clearly established.

Second, while Newton had identified the quantity v^2/r—by an argument broadly similar to that in Chapter 11—he seems to have been in some doubt about what it meant, referring to it on occasion as an 'endeavour *from* the centre' (my italics).

In any event, there was a more serious problem—planetary orbits aren't really circles; they're ellipses.

Newton resumes the attack

It was some ten years later, in 1679, and partly as a result of a letter from Robert Hooke, that Newton took up the problem again.

And he soon showed that if a planet P is subject to a force directed always towards one fixed point, S, then the line SP will sweep out area at a constant rate, i.e. equal areas in equal times.

This result, which holds for a planetary orbit of any shape, was a real breakthrough. For Kepler's second law—if true— could then be explained, instantly, by assuming that the gravitational force on each planet was directed always *towards the Sun*.

Yet this breakthrough, in itself, provided no hint whatsoever about the magnitude of that force, or how it might depend on r.

That came only later still, when Newton finally showed that if, in addition, the orbit is an ellipse, with the Sun S at one focus, then the force must, indeed, be proportional to $1/r^2$.

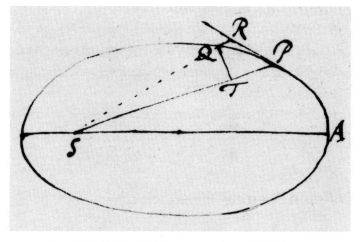

59. A sketch of orbital motion from Newton's unpublished manuscript *De Motu corporum in gyrum* (1684). *S* denotes the Sun.

And, from the point of view of the present book, one of the most interesting things is how he did this.

At a first glance of the manuscripts we are assailed by geometry. But what looks like pure geometry, as in Figure 59, isn't. The planet is first at P and then moves to Q, but, in the end, Newton lets Q become closer and closer to P. In our terms, if δt is the time increase between P and Q, then he eventually lets $\delta t \to 0$.

In other words, the most fundamental idea in the calculus—that of taking a *limit*—is at the heart of what he does, though hardly in the form we would do it today.

And yet, as so often with Newton, all this was done privately, almost secretly, and no one really knew about it until…

Halley's visit to Newton

This famous occasion, probably in August 1684, took place when the astronomer Edmund Halley visited Newton at Cambridge.

By then, the possibility of a gravitational force proportional to $1/r^2$ was a talking point among mathematicians and scientists in the coffee houses of London, and Halley wanted Newton's views on the matter.

According to one of Dr Halley's contemporaries:

> after they had been some time together, the D^r asked him what he thought the Curve would be that would be

described by the Planets supposing the force of attraction towards the Sun to be reciprocal to the square of their distance from it. Sr Isaac replied immediately that it would be an Ellipsis, the Doctor struck with joy & amazement asked him how he knew it, why saith he I have calculated it…

But Newton couldn't find the actual calculation amongst his papers, so promised to send it to Halley as soon as he could.

Halley was, of course, delighted with the prospect, but as his coach clattered back to London he can have had no idea, presumably, that his visit would prompt Newton into eventually producing his great masterpiece on dynamics—the *Principia*.

Nor could he have known, I imagine, that the means for taking Newton's dynamical ideas much, much further—namely the calculus *roughly as we know it today*—was just about to make its first appearance, in a paper by Leibniz.

13

Leibniz's Paper of 1684

From a modern perspective, Leibniz's landmark paper of 1684 is really rather strange.

He leaps straight into a series of general rules for what we call differentiation, but with little explanation of what it all means, and virtually no explanation at all of why it all works.

The first rule concerns the differentiation of a sum, which we would write as

$$\frac{d}{dx}(u + v) = \frac{du}{dx} + \frac{dv}{dx}.$$

This is valuable—though hardly surprising—and we have already used it, several times, in this book. An equivalent rule is given for the difference of two functions.

But Leibniz then gives the rule for differentiating the *product* of two functions of x.

This is far less obvious, and we now know that Leibniz himself got it wrong, at first, in his own early manuscripts.

MENSIS OCTOBRIS A. M DC LXXXIV. 467

NOVA METHODVS PRO MAXIMIS ET MI-
nimis, itemque tangentibus, quæ nec fractas, nec irrati-
onales quantitates moratur, & singulare pro
illis calculi genus, per G.G. L.

SIt axis AX, & curvæ plures, ut VV, WW, YY, ZZ, quarum ordi-
natæ, ad axem normales, VX, WX, YX, ZX, quæ vocentur respective, v, vv, y, z; & ipsa AX abscissa ab axe, vocetur x. Tangentes sint
VB, WC, YD, ZE axi occurrentes respective in punctis B, C, D, E.
Jam recta aliqua pro arbitrio assumta vocetur dx, & recta quæ sit ad
dx, ut v (vel vv, vel y, vel z) est ad VB (vel WC, vel YD, vel ZE) vocetur d v (vel d vv, vel dy vel dz) sive differentia ipsarum v (vel ipsarum vv, aut y, aut z) His positis calculi regulæ erunt tales : __

Sit a quantitas data constans, erit da æqualis o, & d $\overline{ax}$ erit æqu-
a dx: si fit y æqu. v (seu ordinata quævis curvæ YY, æqualis cuivis or-
dinatæ respondenti curvæ VV) erit dy æqu. dv. Jam *Additio & Sub-
tractio:* si sit z -y + vv + x æqu. v, erit d$\overline{z\text{-}\text{-}y}$ + vv + x seu dv, æqu.
dz -dy + dvv + dx. *Multiplicatio,* $\overline{d x v}$ æqu. x dv + v dx, seu posito
y æqu. xv, fiet d y æqu. x d v + v d x. In arbitrio enim est vel formulam,
ut xv, vel compendio pro ea literam, ut y, adhibere. Notandum & x
& d x eodem modo in hoc calculo tractari, ut y & dy, vel aliam literam
indeterminatam cum sua differentiali. Notandum etiam nōn dari
semper regressum a differentiali Æquatione, nisi cum quadam cautio-
ne, de quo alibi. Porro *Divisio*, d—$\frac{v}{y}$ vel (posito z æqu. $\frac{v}{y}$) d z æqu.
$\overline{\pm v dy \mp y dv}$

60. Leibniz's first paper on the calculus, in the *Acta Eruditorum*, 1684.

Differentiating a product

The rule is given by Figure 61, and we may prove it by the same sort of approach we have used previously, in Chapter 5.

Let x increase to $x + \delta x$, and let δu, δv be the consequent increases in u and v.

Then the increase in uv, that is $\delta(uv)$, will be $(u + \delta u)(v + \delta v) - uv$, which is $u.\delta v + v.\delta u + \delta u.\delta v$.

$$\frac{d}{dx}(uv) = u\frac{dv}{dx} + v\frac{du}{dx}$$

61. Differentiating a product.

So, dividing by δx, we get

$$\frac{\delta(uv)}{\delta x} = u\frac{\delta v}{\delta x} + v\frac{\delta u}{\delta x} + \frac{\delta u}{\delta x} \times \delta v.$$

Finally, we let $\delta x \to 0$, so that $\delta u \to 0$ and $\delta v \to 0$ also. The result then follows, from the definition of derivative in Chapter 5, because the final term tends to 0, on account of the factor δv.

When u and v are both positive we may, if we wish, view the result in geometric terms, regarding u and v as the dimensions of a rectangle, and uv as its area (Figure 62).

Plainly, when δu and δv are very small, the small increase in

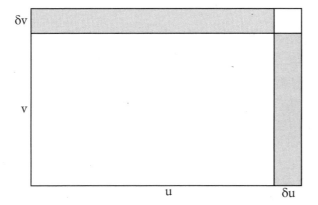

62. Slightly increasing a rectangle.

area is accounted for almost entirely by the area of the two thin (shaded) rectangles, which is $u.\delta v + v.\delta u$, and that is why the rule for differentiating a product takes the form that it does.

Differentiating a ratio

The rule for differentiating the ratio u/v of two functions of x can be deduced in a very similar way.

When x increases to $x + \delta x$, so that u increases to $u + \delta u$ and v to $v + \delta v$, the consequent small increase in u/v is

$$\frac{u + \delta u}{v + \delta v} - \frac{u}{v} = \frac{v.\delta u - u.\delta v}{(v + \delta v)v}.$$

$$\frac{d}{dx}\left(\frac{u}{v}\right) = \frac{v\,\frac{du}{dx} - u\,\frac{dv}{dx}}{v^2}$$

63. Differentiating a ratio.

This, then, is $\delta(u/v)$, and on dividing by δx and letting $\delta x \to 0$ (so that $\delta u \to 0$ and $\delta v \to 0$ also) we obtain the result shown in Figure 63.

This is the last of the general rules in Leibniz's 1684 paper, and we will use it in Chapter 17 to help prove one of the greatest mathematical results of all time.

Differentiating x^n

We claimed earlier, in Chapter 5, that

$$\frac{d}{dx}(x^n) = nx^{n-1},$$

where n is any positive (and constant) whole number, and we can now use Leibniz's product rule to see why this is so.

If we start with our very first, major result,

$$\frac{d}{dx}(x^2) = 2x.$$

we can then differentiate x^3 by regarding it as the product $x^2.x$. Thus, using the product rule:

$$\frac{d}{dx}(x^3) = 2x.x + x^2.1$$
$$= 3x^2.$$

We can then use *this* result, in exactly the same way, to differentiate x^4, obtaining $4x^3$, and if we actually proceed in this way it quickly becomes apparent why the emerging pattern must inevitably continue forever as n increases.

Yet the result is in fact even more general. Leibniz emphasizes in his 1684 paper that the derivative of x^n is nx^{n-1} even when the power n is *fractional* or *negative*.

Note, for example, that $x^{1/2}$ denotes the positive square root of the positive number x, i.e.

$$x^{1/2} = \sqrt{x} \qquad for \quad x > 0,$$

because, by the law of indices, $x^{1/2} \times x^{1/2} = x^1$. And, by similar reasoning,

$$x^{-1} = \frac{1}{x}, \quad x^0 = 1 \quad for \ x \neq 0.$$

And, according to Leibniz, we can differentiate these powers of x by the same general rule.

In the case $n = -1$, for instance, it gives the derivative of $1/x$ as $-1/x^2$, which we already know to be correct from Chapter 5.

Leibniz and the 'infinitely small'

As mentioned earlier, there are no derivations of these results in Leibniz's paper.

And, as the extract in Figure 60 shows, the results are written differently. Leibniz writes the product rule, for example, as

$$d(uv) = v.du + u.dv.$$

Curiously, it is never explained very clearly what quantities like du and dv really are, but in an earlier, unpublished manuscript, from about 1680, Leibniz writes:

$$d(xy) = (x + dx)(y + dy) - xy$$
$$= x.dy + y.dx + dx.dy$$

and says that this

> will be equal to $x.dy + y.dx$ if the quantity $dx.dy$ is omitted, which is infinitely small with respect to the remaining quantities, because dx and dy are supposed infinitely small....

So Leibniz's view seems to be very different from the approach in this book, which is based not on the idea of 'infinitely small', but on the idea of a limit.

A shortest-time problem

Towards the end of his 1684 paper, Leibniz applies his new techniques to one practical problem of real significance.

While he doesn't put it quite like this, we may rephrase the problem using Figure 64. And the question is: how do we get from a point A on the beach to the point B in the sea as quickly as possible?

Now, the shortest *path* from A to B is clearly a straight line, but if we run a lot faster than we swim we may be well advised to take a path more like the one in the figure, involving a greater distance on sand but a shorter distance in the water.

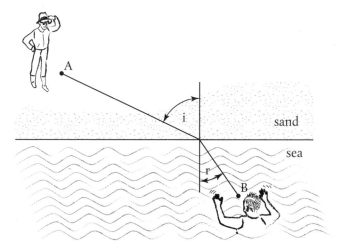

64. A shortest-time problem.

In any event, calculus eventually provides the answer: it turns out that we minimize the time if we choose the angles i and r so that

$$\frac{\sin i}{\sin r} = \frac{c_{sand}}{c_{water}},$$

where c_{sand} is the speed at which we run, and c_{water} the speed at which we swim.

In truth, though, this problem isn't really about running and swimming; it's all about *light*, and that is how Leibniz introduces it in his paper.

When light is refracted, as it passes from one medium to another, the angle of incidence i and the angle of refraction r also satisfy the same equation, with c_{sand} and c_{water} replaced by the speeds of light in the two media.

So calculus shows, then, that when light is refracted at the plane boundary between two media, it travels from one given point to another in the shortest possible time.

And for some people, at least, this always prompts the question: how does light *know* how to take the path of shortest time?

And I have always rather liked the playful (and quantum-mechanical) answer once given by the physicist Richard Feynman: 'It doesn't. It tries them all.'

14

'An Enigma'

The advent of calculus completely transformed mathematics. Yet, at the time, very few mathematicians could understand properly what Newton and Leibniz had done.

Even the great Swiss mathematician John Bernoulli, for instance, described Leibniz's 1684 paper as

> an enigma rather than an explication.

But Bernoulli persisted, and eventually lectured on the subject to—amongst others—the Marquis de l'Hôpital, who went on to publish the first textbook on differential calculus, in 1696.

L'Hôpital's book, *Analyse des infiniment petits pour l'intelligence des lignes courbes*, was enormously influential, and written very much in the notation and spirit of Leibniz's approach to calculus.

One of the earliest calculus textbooks in English, on the other hand, was Charles Hayes' *A Treatise of Fluxions*, published in 1704 (see Figure 65).

The title here is a reference to the way that Newton often thought about some curve in terms of *motion* along it, so that x and y both depend on some time-like variable t. Newton used the term 'fluxion' for the rate at which some variable depends on t, and denoted that by a dot, so that the fluxion of x was $\dot{x}$, and this particular notation for dx/dt is still in use today.

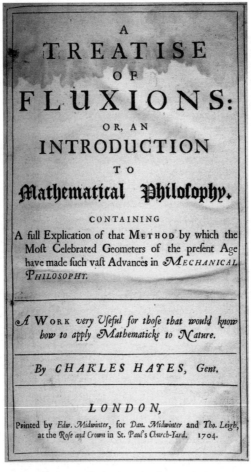

A

TREATISE

OF

FLUXIONS:

OR, AN

INTRODUCTION

TO

𝕸𝖆𝖙𝖍𝖊𝖒𝖆𝖙𝖎𝖈𝖆𝖑 𝕻𝖍𝖎𝖑𝖔𝖘𝖔𝖕𝖍𝖞.

CONTAINING

A full Explication of that METHOD by which the Most Celebrated Geometers of the present Age have made such vast Advances in *MECHANICAL PHILOSOPHY.*

A WORK *very Useful for those that would know how to apply Mathematicks to Nature.*

By *CHARLES HAYES*, Gent.

LONDON,

Printed by *Edw. Midwinter*, for *Dan. Midwinter* and *Tho. Leigh,* at the *Rose and Crown* in St. *Paul's Church-Yard.* 1704.

65. An early textbook on calculus (1704). This particular copy was owned and inscribed by Thomas Foy, a student of Oxford University, in 1709.

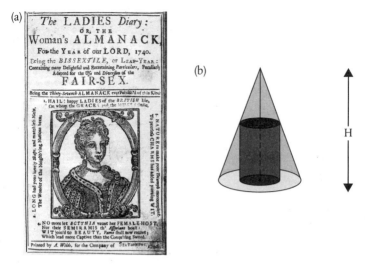

66. (a) The *Ladies Diary*. (b) Mrs. Sidway's problem.

It was through early textbooks such as this, then, that calculus began to spread. And before long it was even reaching some rather unlikely places, including the pages of *The Ladies Diary*, a popular journal of the time which included some mathematical puzzles among its 'Delightful and Entertaining Particulars' (see Figure 66).

And in the 1714 issue, Mrs Barbara Sidway poses a problem involving a circular cylinder inside a cone of given height H.

While it is thinly disguised (in verse) as a gardening question, Mrs Sidway's problem is essentially this: what height should the cylinder be if its volume is to be as large as possible?

The *Diary* eventually received four correct solutions from its readers, and while we cannot be sure of the methods used, calculus certainly gives the right answer: $\frac{1}{3}H$.

Notation, notation...

As we have seen, Leibniz's notation for calculus is still in widespread use today, and one reason for its success is this: while we do not regard dy/dx as a ratio of two quantities dy and dx, it often behaves *as if it is*.

Differentiation

Suppose, for instance, that y is some function of x, and that x itself is some function of another variable—say t. Then we can, if we wish, consider y as a function of t, and then

$$\frac{dy}{dt} = \frac{dy}{dx} \cdot \frac{dx}{dt}.$$

This is a major result in the subject, called the *chain rule*.

A quick way of differentiating $y = (t^2 + 1)^3$ with respect to t, for instance, would be to first set $x = t^2 + 1$, so that $y = x^3$. Then $dy/dx = 3x^2$ and $dx/dt = 2t$, so the chain rule gives $dy/dt = 6t(t^2 + 1)^2$.

One major consequence of the chain rule is

$$\frac{dy}{dx} \cdot \frac{dx}{dy} = 1,$$

and we will use this, shortly, in a rather striking context.

Another piece of Leibniz notation that has stood the test of time is that used when we want to differentiate some function of x *twice*:

$$\frac{d^2 y}{dx^2} \text{ means } \frac{d}{dx}\left(\frac{dy}{dx}\right),$$

and, again, we will use this later in the book.

Integration

As we have seen already, integration can be much more diffi-cult than differentiation, but even here a good notation helps.

And it was, again, Leibniz who introduced the famous 'integral' sign: ∫.

Thus if

$$\frac{dA}{dx} = y,$$

we may write this equivalently as

$$A = \int y \, dx,$$

called 'the integral of y with respect to x' (Figure 67).

The symbol ∫ itself is really just an elongated letter 's' denot-ing 'sum', for A represents the area under the curve of y against x, and that is, indeed (the limit of) the sum of lots of little rectangular areas, each of amount $y \, \delta x$.

Sed ex iis quæ in methodo tangentium expofui, patet efle d, ½ xx=xdx ; ergo contra ½ xx=∫xdx (ut enim poteftates & radices in vulgaribus calculis, fic no-bis fummæ & differentiæ feu ∫ & d, reciprocæ funt.)

67. The first appearance in print of the integral sign ∫, in a paper by Leibniz dated 1686.

So, for example,

$$\int x \, dx = \frac{1}{2}x^2 + \text{constant},$$

and, more generally,

$$\int x^n \, dx = \frac{x^{n+1}}{n+1} + \text{constant} \ \text{ for } n \neq -1$$

Finally, Leibniz's notation helps with one particularly powerful integration technique.

This is integration *by change of variable*, which involves writing x, and therefore y, in terms of some new variable t. The idea is to convert a difficult integration with respect to x into an easier integration with respect to t:

$$\int y \, dx = \int y \frac{dx}{dt} \, dt,$$

and, once again, Leibniz's notation makes the whole procedure seem almost 'natural'—and certainly easy to remember.

Leibniz's emphasis on a good mathematical notation was wholly consistent with his wider philosophical ideas, and he was quite explicit about it, once writing to a friend:

> In symbols one observes an advantage in discovery which is greatest when they express the exact nature of a thing briefly and, as it were, picture it....

15

Who Invented Calculus?

In the Royal Society of London's *Philosophical Transactions* for 1708 there is a largely forgotten paper by the Oxford mathematician John Keill.

Forgotten, that is, save for the following short passage where Keill refers to the calculus

> which Mr. Newton, beyond all doubt, first discovered... though the same Arithmetic was published later by Mr. Leibniz in the *Acta Eruditorum* with changes in the name and method of notation.

When Leibniz eventually saw this, in 1711, he took it as an accusation of plagiarism, and immediately put in an official complaint to the Royal Society, demanding an apology from Keill.

A committee was set up to investigate the matter, but did not uphold Leibniz's complaint.

In retrospect, however, this is hardly surprising, because by that time Newton was President of the Royal Society, and he not only stuffed the committee full of his own supporters, but wrote much of the final report himself.

Newton v. Leibniz

In truth, the question of priority with regard to the calculus had been simmering for years.

We now know that Newton had many of the main results in 1665–6, long before Leibniz had even turned his attention to mathematics.

For much of that time, Cambridge University was closed, because of the plague, and Newton retreated to his family home in Lincolnshire. And for him, at least, this was an extraordinarily creative time.

One striking example is the link between differentiation and the area under a curve, which we would now write (using Leibniz's notation) as

$$\frac{dA}{dx} = y,$$

for this appears—in a different form—in a manuscript dated as early as October 1666, when Newton was only 23.

He wrote a short account of these early results in his *De Analysi* of 1669 (see Figure 68), and in a much more extensive work—*Methodus Fluxionum et Serierum Infinitarum*—two years later, in 1671. And Newton allowed these manuscripts to be seen by a small, select number of contemporary mathematicians.

Somewhat later still, in 1674–6, Leibniz made many of his discoveries in calculus, while working in Paris.

Towards the end of this period, in October 1676, Leibniz visited London on a diplomatic mission, and this lies at the

DE ANALYSI

Per Æquationes Numero Terminorum

INFINITAS.

*Ethodum generalem, quam de Curvarum quanti-
tate per Infinitam terminorum Seriem menfuran-
da, olim excogitaveram, in fequentibus breviter explica-
tam potius quam accuratè demonftratam habes.*

ASI *AB* Curvæ alicujus *AD*, fit
Applicata *BD* perpendicularis : Et
vocetur *AB = x*, *BD = y*, & fint
a, b, c, &c. Quantitates datæ, &
m, n, Numeri Integri. Deinde,

Curvarum Simplicium Quadratura.

REGULA I.

Si $ax^{\frac{m}{n}} = y$; *Erit* $\frac{an}{m+n}x^{\frac{m+n}{n}} = Areæ\ ABD.$

Res Exemplo patebit.

1. Si $x^2\ (= 1x^{\frac{2}{1}}) = y$, hoc eft, $a = 1 = n$, & $m = 2$; Erit $\frac{1}{3}x^3 = ABD.$

A   2. Si

---

68. The first page of Newton's *De Analysi*, as eventually
published in 1711.

heart of the priority dispute. For while he never met Newton, Leibniz was shown, during the visit, some of Newton's early work in manuscript form, including *De Analysi*.

So, while Leibniz was certainly the first to publish—in 1684—his detractors eventually began to ask what he might have gleaned from the London visit and from an occasional—and rather wary—exchange of correspondence with Newton himself.

'The most ... suspicious temper'

It is all too easy to speculate that the whole calculus dispute could have been avoided if Newton had published his works on calculus, in full, earlier.

Why, then, didn't he?

Some scholars have cited the dire state of the book trade, following the Great Fire of London in 1666. Most, however, see the explanation in Newton's own character, which seems to have been extraordinarily introverted and secretive.

According to one contemporary, he had

> the most fearful, cautious and suspicious temper that ever I knew.

And Newton himself admitted to an almost pathological dread of controversy, especially one in print.

In any event, there were other ways, too, in which the whole dispute was slightly absurd.

After all, calculus did not just appear out of nowhere. As we have already seen, it owed much to earlier work by Archimedes, Descartes, Fermat, and Wallis, to say nothing of Isaac Barrow, whom Newton succeeded at Cambridge as Lucasian Professor.

Yet it was Newton and Leibniz who took a whole host of disparate ideas and created the calculus as a coherent subject, centred on the concepts of differentiation, integration, and the fundamental theorem.

And the verdict of most historians of mathematics today is that they did this *independently*, and in really rather different ways.

'They have changed the whole point ... '

The most conspicuous difference between the two approaches lies, perhaps, in the role played by infinite series.

Time and again, Newton used infinite series as an aid to integration, in a way similar to that in Chapter 10.

And here he had what he seemed to view, almost, as a secret weapon—the binomial series:

$$(1+x)^n = 1 + nx + \frac{n(n-1)}{1.2}x^2 + \frac{n(n-1)(n-2)}{1.2.3}x^3 + \cdots$$
$$for\ -1 < x < 1$$

This was already well known when n is a positive whole number, in which case it holds for any x and stops after $n + 1$ terms, because all subsequent coefficients are 0.

But Newton, in one of his earliest and most highly prized mathematical discoveries, realized that it holds *as an infinite series* if *n* is fractional or, even, negative.

Thus, setting $n = -1$ gives an infinite series representation of the function $1/(1 + x)$—in fact, precisely the one we saw in Chapter 10. And setting $n = \frac{1}{2}$, for instance, gives an infinite series for $\sqrt{1+x}$.

And Newton used these ideas so prolifically that it is sometimes difficult, almost, to find him doing what we would call calculus without them.

69. The famous infinite series involving the odd numbers, in Leibniz's own hand, from a letter dated 1676.

For Leibniz, on the other hand, infinite series seem to have been far less central to the subject as a whole, and something of this emerges, even, from a reply of his to the Royal Society's report on the priority dispute:

> They have changed the whole point of the controversy, for in their publication…one finds hardly anything about the differential calculus; instead every other page is made up of what they call infinite series.…

It is a little ironic, then, that one of the most stunning results ever involving infinite series, namely

$$1 - \frac{1}{3} + \frac{1}{5} - \frac{1}{7} + \ldots = \frac{\pi}{4},$$

is usually attributed to Leibniz.

And, as it happens, we are almost ready to see how this extraordinary connection between circles and odd numbers comes about.

Almost.

But not quite…

16

Round in Circles

In mathematics, some functions *oscillate*.

The most well-known examples are sin θ and cos θ, and they have the striking property that they are almost—but not quite—derivatives of one another (Figure 70).

This may possibly come as something of a surprise, because most of us meet sin θ and cos θ for the first time through trigonometry, where θ is one *angle* of a right-angled triangle (Figure 71).

And yet, as we shall see, all these ideas *are* related.

Angles

First, we need to measure any angles which occur not in degrees but in *radians*. These are defined as follows.

Draw a circle, and then move around the circumference by a distance equal to the radius, r.

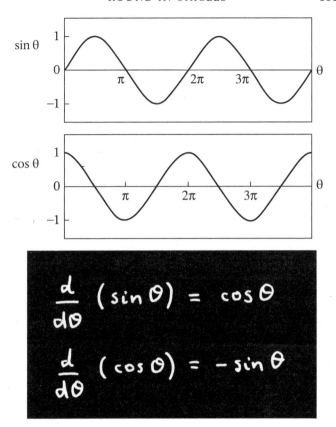

$$\frac{d}{d\theta}(\sin\theta) = \cos\theta$$

$$\frac{d}{d\theta}(\cos\theta) = -\sin\theta$$

70. The functions $\sin\theta$ and $\cos\theta$.

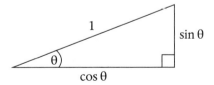

71. A right-angled triangle.

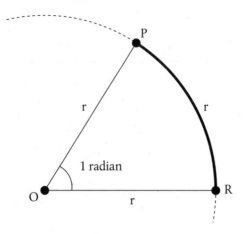

72. Definition of a radian.

This will, by definition, trace out an angle of 1 radian, which is about 57.3 degrees (Figure 72).

And, by the same token,

$$\frac{\pi}{2} \text{radians} = 90 \text{ degrees}$$

because both correspond to going a quarter way round the circumference, which is a distance $\frac{1}{2}\pi r$.

Oscillations

Now draw a circle of radius 1 unit, and imagine a point P which starts at R in Figure 73 and then moves round the circumference of the circle *over and over again*, so that θ, the angle through which P turns, keeps on increasing.

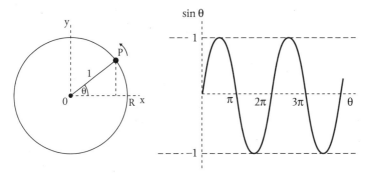

73. Sin θ as an oscillation.

Taking a lead from the elementary geometry of a right-angled triangle, we now *define* cos θ and sin θ, for any number θ, as the x- and y- coordinates, respectively, of the point P.

So, if P starts at R, with $\theta = 0$, the y-coordinate, or sin θ, starts as 0, then goes up to 1 at $\theta = \pi/2$ after one quarter-turn anticlockwise. In subsequent quarter-turns it goes back down to 0, down to −1, and finally back up to 0 again when $\theta = 2\pi$, whereupon the whole business starts again as P makes a second 'orbit' with θ going from 2π to 4π.

And cos θ, the x-coordinate of P, varies with θ in exactly the same way, but out of step by an amount $\pi/2$, as the graphs in Figure 70 show.

This, then, is how− and why—the functions sin θ and cos θ continually *oscillate* as the variable θ is steadily increased.

And, not surprisingly, the most interesting question—from a calculus point of view—concerns the rate at which they do this.

Rates of change

Imagine, now, the point P moving round the circle in such a way that $\theta = t$, where t denotes time, as in Figure 74.

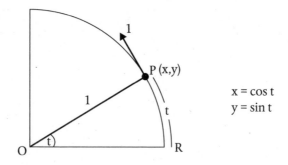

$$x = \cos t$$
$$y = \sin t$$

74. Finding the rates of change.

Then $\dfrac{d}{dt}(\cos t)$ and $\dfrac{d}{dt}(\sin t)$ will simply be dx/dt and dy/dt,

and there is a very simple way of deducing these.

One merit of radian measure—together with a unit radius—is that the distance travelled, PR, is not just proportional to the angle POR—it is actually *equal* to it, and will therefore be t.

So P travels a distance t in time t and therefore goes round and round the circle *at unit speed*. Its velocity at any moment is therefore 1, directed along the tangent.

And because the tangent is perpendicular to the radius OP, this direction of motion makes an angle t with the y-axis (Figure 75).

Velocity

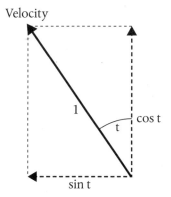

75. The velocity components.

Now, moving with speed 1 in the direction shown is equivalent to moving in the negative x-direction with speed sin t, at the same time as moving in the y-direction with speed cos t. So $dx/dt = -\sin t$ and $dy/dt = \cos t$.

And that is why

$$\frac{d}{dt}(\sin t) = \cos t$$

$$\frac{d}{dt}(\cos t) = -\sin t,$$

as we claimed at the beginning of the chapter.

As we shall see, these ideas are crucial for virtually any physical problem involving oscillations.

But, more surprisingly, perhaps, they also provide the key to unlocking the mysteries of the Leibniz series.

17

Pi and the Odd Numbers

> You have discovered a very remarkable property of the circle,
> which will forever be famous among geometers.
> Christiaan Huygens to Leibniz, in a letter of 1674

We are now—at last—in a position to shed light on one of the
most extraordinary results in the whole of mathematics, linking π and the odd numbers.

$$\frac{\pi}{4} = 1 - \frac{1}{3} + \frac{1}{5} - \frac{1}{7} + \cdots$$

76. The Leibniz series.

The history of this result is a little curious. It was first published by Leibniz, without any derivation or proof, in the *Acta
Eruditorum* for 1682, but he had discovered it much earlier, in
about 1674, while working in Paris.

It is likely, however, that the Scottish mathematician James
Gregory knew the result a few years earlier still.

What seems more certain is that the result was known to
mathematicians in Kerala, India much earlier, and possibly *three
centuries* before Leibniz or Gregory, for it is now often attributed

to Madhava, who founded the Kerala school. Their methods, however, were rather different, and more highly geometrical.

In any event, if we are to use calculus to see, at last, how π and the odd numbers are connected, we are going to need virtually all of the most important ideas we have seen so far.

It will therefore be helpful, I think, to split the argument into several stages.

In search of $\pi/4$...

Consider two numbers x and θ related in the following way:

$$x = \frac{\sin\theta}{\cos\theta},$$

for values of θ between 0 and $\pi/4$ (Figure 77).

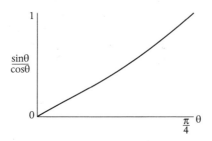

77. The function $\sin\theta/\cos\theta$.

Note, first, that

$$x = 0 \quad \text{when} \quad \theta = 0,$$

and that as we increase θ the value of $x = \sin\theta/\cos\theta$ gradually increases until

$$x = 1 \quad \text{when} \quad \theta = \frac{\pi}{4}.$$

The reason for this is that an angle of $\pi/4$ radians corresponds to 45°, and the right-angled triangle defining sin θ and cos θ is then isosceles, so that the two shorter sides are equal (Figure 78).

This, then, is how $\pi/4$ is going to enter our argument; it is the special value of θ which makes $x = \sin\theta/\cos\theta$ equal to 1.

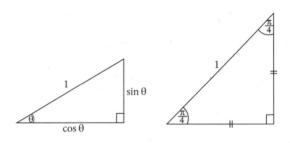

78. Why $x = 1$ when $\theta = \pi/4$.

In search of an infinite series...

This is where calculus really kicks in, and we can split the development into six small steps.

Step 1. Differentiate

$$x = \frac{\sin\theta}{\cos\theta},$$

using Figure 70 *and* Leibniz's rule for differentiating a ratio (Figure 63) to obtain

$$\frac{dx}{d\theta} = \frac{\cos\theta.\cos\theta - \sin\theta.(-\sin\theta)}{(\cos\theta)^2}.$$

Step 2. Use $x = \sin\theta/\cos\theta$ to rewrite the above right-hand side in terms of x itself:

$$\frac{dx}{d\theta} = 1 + x^2.$$

Step 3. Use Leibniz's chain rule (Chapter 14) to rewrite this as

$$\frac{d\theta}{dx} = \frac{1}{1 + x^2},$$

so that we now think of θ as a function of x, rather than the other way round.

Step 4. Now re-write the right-hand side *as an infinite series*, by replacing x by x^2 in the infinite series from Chapter 10:

$$\frac{d\theta}{dx} = 1 - x^2 + x^4 - x^6 + \ldots$$

This step will be valid if $x^2 < 1$.

Step 5. Now use calculus again, this time to integrate with respect to x, in a similar way to that in Chapter 10.

This gives

$$\theta = x - \frac{x^3}{3} + \frac{x^5}{5} - \frac{x^7}{7} + \ldots,$$

the constant of integration being 0, because $x = 0$ when $\theta = 0$.

Step 6. Finally, recall that $x = 1$ when $\theta = \pi/4$, as we observed at the beginning.

On substituting these values in, we obtain Leibniz's famous result, linking π and the odd numbers:

$$\frac{\pi}{4} = 1 - \frac{1}{3} + \frac{1}{5} - \frac{1}{7} + \cdots.$$

(Figure 79).

 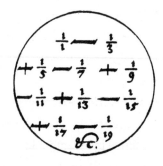

79. Illustrations from Leibniz's paper of 1682.

Before we leave this chapter, however, I ought to make a number of remarks.

First, there was a certain amount of living dangerously in the very last step. Step 4 was valid for x^2 less than 1, yet we set x actually *equal* to 1 in step 6. This can be justified, but only by a more rigorous and technically demanding argument.

Second, this whole approach is not exactly what Leibniz did—his treatment was rather more geometrical, and he explained it in a letter sent (indirectly) to Newton in August 1676.

Thirdly—and somewhat incidentally—Newton immediately fired back a similar-looking series of his own,

$$\frac{\pi}{2\sqrt{2}} = 1 + \frac{1}{3} - \frac{1}{5} - \frac{1}{7} + \frac{1}{9} + \frac{1}{11} - \cdots,$$

though he was at pains to point out that

> y^e signes of y^e series ... are rightly put ... it being a different series from y^t of M. Leibnitz.

Finally, we need to face the fact—pointed out somewhat sarcastically by Newton—that the Leibniz series is hopeless as a practical device for actually calculating π, because it converges so slowly.

Even after 300 terms, for instance, it manages to estimate π less accurately than the well-known approximation 22/7, obtained by Archimedes roughly 2000 years earlier!

But while there are plenty of other infinite series which converge faster to some number involving π, none of them comes close—in my opinion—to the breathtaking elegance and simplicity of the Leibniz series.

18

Calculus under Attack

Leibniz died in 1716.

It was a strange end for one of the greatest mathematicians and philosophers that the world has ever known, for no one attended his funeral except a few friends and his secretary.

And just over ten years later, Newton, too, was gone.

Now, therefore, it fell to others to take calculus further, and one serious matter concerned the logical foundations of the whole subject.

These were brought into particularly sharp relief in 1734 by an essay entitled *The Analyst, or a Discourse Addressed to an Infidel Mathematician* (Figure 80).

The author was George Berkeley, Bishop-elect of Cloyne in Ireland, and the 'infidel mathematician' in question is generally thought to be Edmund Halley, who was a well-known agnostic.

Berkeley was essentially challenging any mathematicians who viewed religion as having shaky foundations to put their own house in order first.

He questioned, even, whether some of the concepts used in calculus actually *exist*. In the most well-known and oft-quoted

THE
ANALYST;
OR, A
DISCOURSE

Addressed to an

Infidel MATHEMATICIAN.

WHEREIN

It is examined whether the Object, Principles, and Inferences of the modern Analysis are more distinctly conceived, or more evidently deduced, than Religious Mysteries and Points of Faith.

By the AUTHOR of *The Minute Philosopher.*

First cast out the beam out of thine own Eye; and then shalt thou see clearly to cast out the mote out of thy brother's eye. S. Matt. c. vii. v. 5.

LONDON:
Printed for J. TONSON in the *Strand.* 1734.

80. Berkeley's essay *The Analyst* (1734).

part of *The Analyst*, for instance, he directs his sarcasm at Newton's whole idea of fluxions, which involve consideration of what Newton calls 'evanescent increments'.

Berkeley is scathing:

> And what are these same evanescent Increments? They are neither finite Quantities nor Quantities infinitely small, nor yet nothing. May we not call them the Ghosts of departed Quantities?

Arguably, however, Berkeley's most trenchant criticism is directed at the actual *reasoning* used in calculus.

Rate-of-change revisited

To see something of Berkeley's objections, consider again what we actually *do*, algebraically, when we differentiate even something as simple as $y = x^2$.

First, we increase x to $x + h$, say, so that the consequent increase in y is $(x + h)^2 - x^2 = 2hx + h^2$.

We then divide one increase by the other:

$$\frac{2hx + h^2}{h}, \quad (i)$$

and cancel the factor of h to get

$$2x + h. \quad (ii)$$

Finally, we omit (or 'blot out', as Newton liked to say) the last term to obtain

$$2x \quad (iii)$$

as the derivative of x^2.

But Berkeley would immediately ask: *is h zero or not?* For, if h is zero, then stage (i) is not allowed, because you can't divide by zero. But if h isn't zero, then some kind of error is made in passing from (ii) to (iii).

In Berkeley's eyes we seem, as it were, to be having our cake and eating it:

> All which seems a most inconsistent way of arguing, and such as would not be allowed of in Divinity.

And Berkeley was just as unimpressed by the standard justification at the time for what is going on, namely that h is 'infinitely small':

> Now to conceive a quantity infinitely small, that is, infinitely less than any sensible or imaginable quantity ... is, I confess, above my capacity.

He just didn't believe that such things exist.

Did Newton or Leibniz really believe in the 'infinitely small'?

We have already seen Leibniz, in about 1680, invoking the 'infinitely small' (Chapter 13).

And Newton, too, used the idea in his early work on calculus, though he was evidently uncomfortable with it. This comes across quite clearly in a manuscript from 1665 where he remarks that the mathematical operations he is performing cannot be allowed

> unlesse infinite littlenesse may bee considered geometrically.

Yet, as time went on, both men seem to have moved away from the idea.

An early (1710) translation of Newton's *De Quadratura*, for example, begins:

> I don't here consider Mathematical Quantities as composed of Parts extreamly small, but as generated by a continual motion…

while Leibniz writes, in a letter of 1706:

> Philosophically speaking, I no more believe in infinitely small quantities than in infinitely great ones…I consider both as fictions of the mind for succinct ways of speaking….

In this way, Newton and Leibniz were well aware that their deductive arguments lacked the rigour of the ancient Greeks, with their brilliant (but often cumbersome) proofs-by-contradiction.

But for Leibniz, in particular, such rigour was not the top priority. The key question was, rather, 'Does calculus give correct results?', and, even more importantly, 'Does it facilitate the *discovery* of new ones?'

Limits

I should like to end this chapter by returning for a moment to the steps involved in the differentiation of $y = x^2$.

For we may well claim, of course, that we do not get from $2x + h$ to $2x$ by 'setting $h = 0$' but, instead, by 'taking the *limit* of $2x + h$ as h *tends* to 0'.

I imagine, however, that Berkeley would immediately ask what we really *mean* by this, exactly.

And, in the event, it took mathematicians a very long time indeed to put the whole idea of 'limit' on a rigorous footing—as we will see later.

In the meantime, calculus just raced ahead, sometimes at almost breakneck speed.

Because it *worked*.

19

Differential Equations

> It appears clear to me...that foreign Mathematicians have,
> of late, been able to push their Researches farther, in many
> particulars, than Sir Isaac Newton and his followers here,
> have done.
>
> British mathematician Thomas Simpson, writing in 1757

The next towering figure in our story is Leonhard Euler (1707–83).

Euler was Swiss, and studied with John Bernoulli, but then spent most of his mathematical career in Berlin and St. Petersburg.

And according to a contemporary:

> Leonhard Euler is not, like the great algebraists usually
> are, of sinister character and clumsy behaviour, but cheer-
> ful and lively.

He was also one of the most prolific mathematicians who ever lived, and the St. Petersburg Academy was still publishing his legacy of scientific papers some fifty years after his death.

Several of his most important contributions were to dynamics, where by building on Newton's groundbreaking

81. Leonhard Euler.

work he helped lay the foundations for an approach to the subject which is still in widespread use today.

And one key idea is to first formulate the physical problem in terms of *differential equations*.

A differential equation is one in which we are told something about the *rate* at which some quantity is changing. Our task is then to determine how the quantity itself changes with time.

And to illustrate this, we now turn to one of the oldest subjects of scientific enquiry.

The simple pendulum

In its most primitive form, a simple pendulum is just a mass suspended from a fixed point by a length of string.

And it turns out that *small* oscillations of such a pendulum are then governed by the differential equation in Figure 82.

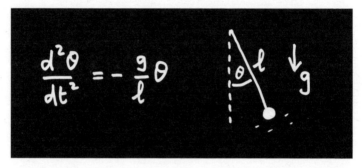

82. The differential equation for small oscillations of a simple pendulum.

Here θ is the angle (in radians) between the pendulum and the vertical at time t, while l denotes the length of the pendulum and g the acceleration due to gravity (9.81 m s^{-2}).

The equation itself is essentially just a statement of the fundamental law of motion: *force = mass × acceleration*, in a direction perpendicular to the string.

Without going into all the details, we should note that the right-hand side is *proportional to θ*, and comes from the force due to gravity. The minus sign arises because that force is always trying to push the pendulum *back* towards the downward-hanging $\theta = 0$ state.

The left-hand side, on the other hand, comes from the acceleration, and $d^2\theta/dt^2$ denotes the second derivative of θ, as explained in Chapter 14.

And our task now is to solve this equation to find how the angle θ depends on time t.

The nature of the problem

We are faced, then, with

$$\frac{d^2\theta}{dt^2} = -\frac{g}{l}\theta,$$

and a perfectly reasonable first reaction would be: 'Integrate twice with respect to t.'

But there's a problem.

And it's rather serious.

The problem is not that the right-hand side is some tremendously awkward function of t, making integration with respect to t difficult.

The problem is that the right-hand side isn't given in terms of t *at all*; it's given in terms of θ, and we have no idea at the outset how θ depends on t, because *that is what we are trying to find out*.

This is absolutely typical of differential equations, and why they often call for a great deal of ingenuity.

A solution

As it happens, in this particular case, it isn't quite true that we have no idea how θ depends on t; we are expecting the pendulum to *oscillate*.

Now, we saw in Chapter 16 that the functions $\cos t$ and $\sin t$ are oscillatory, so let us allow ourselves a bit more leeway and try a solution

$$\theta = A\cos\omega t,$$

where A and ω are both constant. A will then measure the size of the oscillations (assumed small) and ω will measure how rapidly they occur.

A slight generalization (by Leibniz's chain rule) of the results in Figure 70 then gives

$$\frac{d}{dt}(\cos\omega t) = -\omega\sin\omega t$$

$$\frac{d}{dt}(\sin\omega t) = \omega\cos\omega t.$$

So when we differentiate $\theta = A\cos\omega t$ *twice* we get

$$\frac{d^2\theta}{dt^2} = -A\omega^2\cos\omega t$$
$$= -\omega^2\theta.$$

Suddenly, then, we see that θ will be a solution of the original differential equation if $\omega = \sqrt{g/l}$, in which case

$$\theta = A\cos\left(\sqrt{\frac{g}{l}}t\right).$$

And this is, in fact, *the* solution of the problem if the pendulum starts, at $t = 0$, from a stationary position making a small angle A with the vertical.

The oscillation period

At this point, the obvious question is: how long does it take for each complete oscillation?

And we can answer this quite simply, because we know from Chapter 16 that the function $\cos x$ performs one complete oscillation whenever x increases by 2π.

The time for one complete oscillation of the pendulum must therefore be

$$T = 2\pi\sqrt{\frac{l}{g}}.$$

This is one of the oldest and most well-known formulae in the whole of physics, and we have just seen how it follows directly from the law *force = mass × acceleration*, together with a bit of calculus.

Notably, the oscillation period doesn't depend on the constant A, so provided the oscillations are small, it doesn't matter exactly how small they are.

But the most striking feature, surely, is that T is *proportional to the square root of the length l.*

This was discovered by Galileo, in around 1609, in one of his most famous experiments. And we can, if we wish, follow (loosely) in his footsteps.

To do this, just set a pendulum swinging, and count every time it performs *half* a complete oscillation, by reaching one end or other of its swing.

Next, while still counting, shorten the string by a factor of 4.

When you set the pendulum swinging again it should then perform—quite convincingly—a complete to-and-fro oscillation, in time with your count.

20

Calculus and the Electric Guitar

While differential equations are the key to understanding the physical world, they are often of a rather different kind from anything we have met so far.

This is simply because, all too often, the quantity we are trying to determine depends on *more than one variable*.

If you pluck a guitar string, for instance, the string displacement y plainly depends not only on time t but on the distance x from one end (Figure 83).

83. Vibrations of a guitar string.

So y is a function of *two* variables, t and x, and a more sophisticated form of calculus is therefore needed, involving things called *partial derivatives*:

$$\frac{\partial y}{\partial t} \quad \text{and} \quad \frac{\partial y}{\partial x}.$$

The first of these is simply the rate of change of y with t at a fixed value of x, and it is therefore the vibration velocity of the string at that particular point.

In a similar way, $\partial y/\partial x$ is the rate of change of y with x at a fixed time t, so that it represents the slope of the string at that particular moment, as if we were taking a 'snapshot'.

And the slightly different notation '∂'—a sort of curly 'd'—is simply to remind us that we are now differentiating a function of more than one variable.

The wave equation

Suppose, then, that a guitar string has tension T and mass per unit length ρ. It turns out that the displacement y is governed by a *partial differential equation* (Figure 84).

Here, $\partial^2 y/\partial t^2$ is the acceleration of a small bit of the string, and the right-hand side is the force (per unit mass) causing it.

$$\frac{\partial^2 y}{\partial t^2} = \frac{T}{\rho} \frac{\partial^2 y}{\partial x^2}$$

84. The partial differential equation for a vibrating string.

To see why that force takes the form that it does, imagine taking a snapshot of that small bit of the string. If $\partial^2 y/\partial x^2 > 0$, then the slope of the string $\partial y/\partial x$ is increasing with x at that moment, so that particular bit of the string curves slightly 'upwards' (Figure 85).

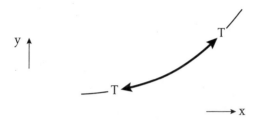

85.   Forces on a tiny portion of the string.

The upward pull from the right-hand portion of the string is then slightly greater than the downward pull from the left-hand portion, resulting in a net force upward, i.e. in the positive y-direction.

In short, it is the curvature of that little bit of the string that gives rise to the net force that we see in the partial differential equation.

That equation itself, known as *the wave equation*, was first derived, and solved, by Jean le Rond D'Alembert, in 1747. And the most striking feature of his solution is that it involves *travelling waves*. These are disturbances which travel along the string, in the x-direction, without change of shape (Figure 86).

Moreover, the speed at which they travel is $\sqrt{T/\rho}$, so the greater the tension T in the string, the faster they go. In fact, they travel so fast on a guitar string that it's almost impossible

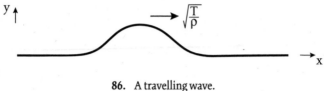

86. A travelling wave.

to see them, but they can be seen quite clearly on a slack washing line, for instance, where T/ρ is typically so much smaller.

Vibrating strings

In order to understand the *sounds* of a guitar string, however, we need to examine some rather different solutions.

Suppose, then, that the string is of length l, and extends between $x = 0$ and $x = l$, where it is fixed, so that $y = 0$ there.

The simplest solution of the partial differential equation

$$\frac{\partial^2 y}{\partial t^2} = \frac{T}{\rho} \frac{\partial^2 y}{\partial x^2}$$

then turns out to be of the form

$$y = A \sin\frac{\pi x}{l} \cos\omega t,$$

where ω is a constant which we will discuss shortly.

The whole string therefore vibrates with a single period $2\pi/\omega$, but different parts of the string, corresponding to different values of x, vibrate by different amounts (Figure 87).

87. The fundamental mode.

In particular, y is always 0 at the two ends $x = 0$ and $x = l$, as required, because $\sin 0 = \sin \pi = 0$ (see Figure 70).

This motion, in which all parts of the string are moving in the same direction at any given moment, is called the 'fundamental' mode.

And the *frequency* of this mode—i.e. the number of vibrations per unit time—is

$$\frac{\omega}{2\pi} = \frac{1}{2l}\sqrt{\frac{T}{\rho}}.$$

This emerges at once if we substitute the expression for y into the differential equation itself, in much the same way that we saw with the pendulum problem in Chapter 19.

For any particular guitar string, the tension T and density ρ tend to be fixed, so the feature of most interest here is that the vibration frequency is proportional to $1/l$.

This is why pressing down on a fret, and therefore shortening the string, produces a higher note.

In particular, pressing down on the 12th fret *halves* the length of the string, l, and therefore doubles the fundamental frequency, which is why the resulting note sounds an octave higher than the open string.

As it happens, however, the fundamental mode is only the first of a whole sequence (Figure 88).

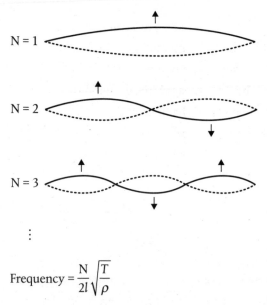

$$\text{Frequency} = \frac{N}{2l}\sqrt{\frac{T}{\rho}}$$

88. Modes of vibration.

And, most strikingly, the vibration frequency of each mode is a whole-number multiple, N, of the fundamental frequency.

Again, this emerges from the differential equation itself, though conditions at the two ends also play a crucial role. This is because, for these higher modes, y is proportional to $\sin N\pi x/l$, and this is only 0 at the right-hand end, $x = l$, if N is a whole number (see Figure 70).

In particular, then, the $N = 2$ mode—in which the two halves of the string move in opposite directions at any given moment—vibrates at twice the frequency of the fundamental, and therefore sounds an octave higher.

In practice, when we pluck a guitar string, the response is typically a complicated mixture of *all* these different modes. And while the fundamental, $N = 1$, tends to dominate, it is possible to give more emphasis to the higher harmonics by plucking the string near to one end, and this is why the resulting note then sounds harsher, and less well-rounded.

In addition, there are various more sophisticated playing techniques—well known to rock guitarists—for suppressing some modes of vibration and highlighting others. And several of these 'tricks' exploit the fact that the higher modes have *nodes*, or points of no motion, at select places along the string.

Often, then, it is by artificially creating a suitable node that you get the particular mode you want—if you're lucky.

21

The Best of all Possible Worlds?

> Nature operates by the simplest and most expeditious ways
> and means.
>
> Pierre de Fermat, 1662

The idea that we might be living in 'the best of all possible worlds' is one of Leibniz's most controversial contributions to philosophy, and in 1759 it was famously ridiculed in Voltaire's satirical novel *Candide*.

Yet the possibility that our world might be optimal *in some way* was, at the time, acquiring a certain scientific credibility.

As early as 1662, for instance, Fermat had proposed that light always travels from one given point to another in such a way as to take the *least time*. And, as we saw in Chapter 13, Leibniz himself used his brand new differential calculus to show that light does indeed behave in this way when refracted at a plane boundary (Figure 89).

It is true that critics had been quick to point out some exceptions to the rule—including, for example, reflection in a concave spherical mirror—but this had not stopped ideas of this general kind being explored, and by the middle of the 18th century they had entered mechanics as well as optics.

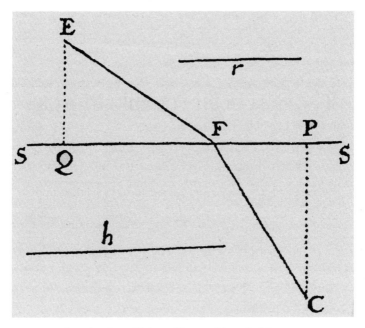

89. The refraction of light, as illustrated in Leibniz's 1684 paper.

And, not surprisingly, all this helped trigger a renewed mathematical interest in problems of *optimization*.

Yet to understand something of this, we need now to broaden our ideas well beyond those of Chapter 6.

Optimization extended

First, the quantity that we want to maximize or minimize may depend on more than one variable.

To illustrate this, I would like to consider a specific problem, even though it will be scarcely more credible, I fear, than the farmer-and-his-field of Chapter 6.

Imagine nonetheless, if you will, that we want to make a bookcase of given volume V, with two shelves, using as little material as possible (Figure 90).

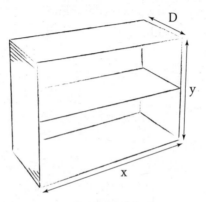

90. A two-shelf bookcase.

With width x, height y, and depth D, the total surface area will be $A = xy + 2yD + 3xD$. And if we use the given volume $V = xyD$ to eliminate D, say, then

$$A = xy + \frac{2V}{x} + \frac{3V}{y}.$$

So, if we wish to minimize A, in order to use as little material as possible, we must minimize a function of *two* independent variables, x and y.

And we can do this by calculating the two partial derivatives:

$$\frac{\partial A}{\partial x} = y - \frac{2V}{x^2},$$

$$\frac{\partial A}{\partial y} = x - \frac{3V}{y^2},$$

and setting *both* equal to 0. This gives two equations for the two unknowns x and y, and on combining those with $V = xyD$ we learn that the width, height, and depth must be in the proportion 2:3:1.

Now, as it happens, this *is* the solution to our problem, but the situation in general is rather more complicated.

For if z is some function of two variables, x and y, we can think of it geometrically as a surface. And the three functions in Figure 91 all have both partial derivatives 0 at $x = 0, y = 0$. Yet the first has a minimum there, the second a maximum, and the third neither, for the origin of coordinates in that case is a 'saddle-point'.

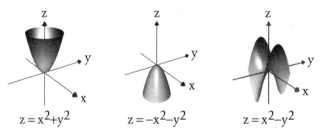

$$z = x^2 + y^2 \qquad z = -x^2 - y^2 \qquad z = x^2 - y^2$$

91. Some functions of two variables.

So, as with the optimization problems of Chapter 6, setting derivatives equal to 0 is only part of the story.

The calculus of variations

There is an even more demanding type of problem, where the quantity that we are trying to maximize or minimize depends on a whole *curve* or *surface*.

The most famous example is, perhaps, the 'brachistochrone' problem, posed by John Bernoulli in 1696. The question is: which curve, between two given points A and B, allows an object to descend under gravity *in the shortest possible time?*

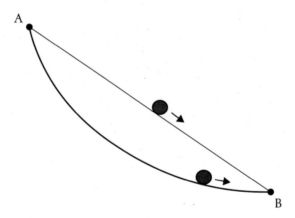

92. The brachistochrone problem.

Galileo had shown, much earlier, that the shortest *path*—a straight line—is not the answer, but had mistakenly claimed that the real answer was the arc of a circle.

Bernoulli showed, however, that the real answer is an upside-down *cycloid* (Figure 92), a cycloid being the curve traced out by a point on the rim of a wheel rolling along a horizontal surface.

In general, problems of this kind call for a sophisticated branch of the subject called the *calculus of variations*, developed in the 18th century by Euler and by Joseph-Louis Lagrange (1736–1813). And the outcome is typically a differential equation for the curve or surface that has the desired maximal or minimal property.

Imagine, for instance, two circular hoops with a soap film extending between them (Figure 93). The film will try to settle in such a way that its surface area is as small as possible, in order to minimize its surface energy.

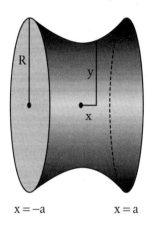

x = −a   x = a

93. A soap film between two hoops.

And, according to the calculus of variations, the differential equation for its radius y is

$$y\frac{d^2y}{dx^2} - \left(\frac{dy}{dx}\right)^2 = 1.$$

The mathematical problem, then, is to solve this equation subject to the boundary conditions that $y = R$ when $x = -a$ and when $x = a$.

As it happens, however, the most intriguing feature in this case is not the solution itself, but the way in which there is no solution *at all* if $a/R > 0.6627$, that is if the two hoops are further apart than about $\frac{2}{3}$ of their diameter.

And if, in an actual experiment, we gradually increase the separation distance beyond this critical value, the whole film suddenly collapses—for no apparent reason—into two flat, circular films, one on each hoop.

22

The Mysterious Number e

In calculus, one particular number stands out as 'special':

$$e = 1 + 1 + \frac{1}{1 \times 2} + \frac{1}{1 \times 2 \times 3} + \frac{1}{1 \times 2 \times 3 \times 4} + \cdots$$
$$\approx 2 \cdot 718.$$

And to see how this number arises we start with a rather unlikely subject—the spread of disease.

Exponential growth

In the early stages of an epidemic, the number of cases typically doubles in some given time—say a few days.

So, if we use this 'doubling time' as our unit of time, the number of cases at time $t = 0, 1, 2, 3, 4, \ldots$ will be 1, 2, 4, 8, 16,..., i.e. 2^t. This is so-called *exponential growth*, and it is a direct mathematical consequence of the very natural assumption (at least in the early stages) that the rate of infection will be proportional to the number of people who have the disease already.

And this result has its direct counterpart in calculus, where the function $y = 2^t$ is defined for all t, and not simply when t is a whole number.

For the rate of change of $y = 2^t$ turns out to be *proportional to 2^t itself*.

The function e^t

More notably still, there is a slightly larger number e such that e^t is actually *equal* to its own derivative:

$$\frac{d}{dt}(e^t) = e^t.$$

And this is, arguably, the key property that singles out e as a special number in calculus.

While

$$e^o = 1,$$

in accord with Chapter 13, the function $y = e^t$ increases rapidly with t, as Figure 94 shows.

The simplest way of actually calculating the number e is, perhaps, to represent e^t as an infinite series:

$$e^t = 1 + t + \frac{t^2}{1 \times 2} + \frac{t^3}{1 \times 2 \times 3} + \frac{t^4}{1 \times 2 \times 3 \times 4} + \cdots$$

And it is easy to check that this is, indeed, the correct one. For if we differentiate, we get

$$\frac{d}{dt}(e^t) = 0 + 1 + \frac{2t}{1 \times 2} + \frac{3t^2}{1 \times 2 \times 3} + \frac{4t^3}{1 \times 2 \times 3 \times 4} + \cdots$$

and after some obvious cancellation we realize that the right-hand side is *the original series itself*, i.e., e^t.

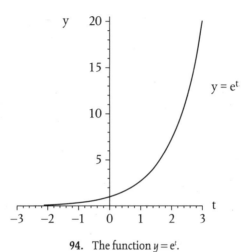

94. The function $y = e^t$.

It also satisfies the requirement that $e^0 = 1$, because all terms but the first are then 0.

This series turns out to be convergent for all t, and by setting $t = 1$ we finally obtain the series for e at the start of this chapter:

$$e = 1 + 1 + \frac{1}{1 \times 2} + \frac{1}{1 \times 2 \times 3} + \frac{1}{1 \times 2 \times 3 \times 4} + \cdots$$

n	sum of the first n terms
1	1
2	2
3	2·5
4	2·666 ...
5	2·7083 ...
6	2·7166 ...
7	2·71805 ...
8	2·71825 ...

Convergence is rapid, and the most well-known approximation to e, namely 2·718, emerges after only 7 terms of the series.

e and Euler

The number e has a complicated history, but it first came to prominence in Euler's classic *Introductio in analysin infinitorum* of 1748.

Euler introduced it, however, in a different way, which we would now write as:

$$e = \lim_{n \to \infty} \left(1 + \frac{1}{n} \right)^n.$$

This limit is intriguing, because any *fixed* number greater than 1, raised to an ever-increasing power, would tend to infinity. But here, as the power goes up, the number being raised to that power goes *down*, and edges closer and closer to 1, in just such a way that a finite limit results.

e and gambling

Suppose that the chances of winning the jackpot on a slot machine are 1 in 100, and we play it 100 times.

What is the probability of winning?

Well, the probability of losing every single time is $\left(1-\frac{1}{100}\right)^{100}$, which is very close to e^{-1}, i.e. $1/e$, and therefore about 37%.

So the probability of winning is about 63%.

e and logarithms

Sharp-eyed readers may have noticed that there was one exception to the rule in Chapter 14 for integrating any power of x.

The exceptional case is x^{-1}, or $1/x$, and, somewhat curiously, this has a completely different integral involving the *logarithm* of x *to base* e:

$$\int \frac{1}{x} dx = \log_e x + \text{constant.}$$

e and the search for happiness

When searching for a partner, the best strategy, apparently, is to reject the first $1/e$ possibilities—that is, the first 37%—and then settle down with the first new possibility who is better than any of the first 37%.

I say 'apparently', because I haven't actually tried it.

23

How to Make a Series

One of Euler's many contributions to calculus was a subtle change of viewpoint as to what the subject is really all about.

In its early stages, calculus was viewed in a very geometric way, and seen as being all about curves, and their various properties. In the 18th century, however, a more algebraic viewpoint began to emerge, with Euler and others seeing calculus as being all about *functions*.

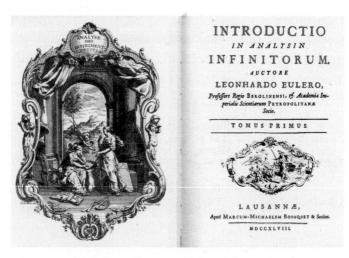

95. Euler's *Introductio in Analysin Infinitorum* of 1748.

And it was Euler who introduced the notation

$$y = f(x),$$

now almost universal, to denote that y is some function of x.

The modern view of the function itself, f, is—as we have seen—just some rule that assigns to each value of x a definite and *unique* value of y, such as $f(x) = x^2$ or $f(x) = \sin x$.

It is sometimes convenient, too, to use a dash to denote a derivative. Thus,

$$f'(x) = \frac{dy}{dx}, \ f''(x) = \frac{d^2 y}{dx^2},$$

and so on (see Chapter 14).

But my real reason for introducing the notation at this stage is in connection with infinite series.

I am well aware, for instance, that when I produced a series for e^t in Chapter 22, it was rather like pulling a rabbit out of a hat.

And in Figure 96 we see Newton obtaining infinite series for $\sin \theta$ and $\cos \theta$, in 1669, but by a brilliant ad hoc method that would be difficult to implement more generally.

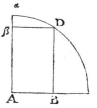

Si ex dato arcu $_\alpha$D Sinus AB defideratur ; æquationis $z = x + \frac{1}{6}x^3 + \frac{3}{40}x^5 + \frac{5}{112}x^7$, &c. fupra inventæ, (pofito nempe $\overline{AB} = x$, $_\alpha D = z$, & $A_\alpha = 1$,) radix extracta erit $x = z - \frac{1}{6}z^3 + \frac{1}{120}z^5 - \frac{1}{5040}z^7 + \frac{1}{362880}z^9$, &c.

Et præterea fi Cofinum Aβ ex ifto arcu dato cupis, fac Aβ ($= \sqrt{1-xx}$) $= 1 - \frac{1}{2}z^2 + \frac{1}{24}z^4 - \frac{1}{720}z^6 + \frac{1}{40320}z^8 - \frac{1}{3628800}z^{10}$, &c.

96. Infinite series for $\sin \theta$ and $\cos \theta$ in Newton's *De Analysi* of 1669 (published 1711). His z is our θ, and his x our $\sin \theta$.

So it is only natural to ask whether there is some easier, more routine way of representing a given function as an infinite series.

Taylor series

Suppose, then, that we want to write some function $f(x)$ in the form

$$f(x) = A + Bx + Cx^2 + Dx^3 + \cdots.$$

The obvious question is: how do we determine the constants $A, B, C, \ldots$?

And, somewhat remarkably, there is a very simple way of doing this. We just differentiate, *repeatedly*, with respect to x, term by term:

$$f'(x) = B + 2Cx + 3Dx^2 + \cdots$$

$$f''(x) = 2C + 2.3Dx + \cdots \quad .$$

$$f'''(x) = \quad 2.3D + \ldots$$

Finally, we set $x = 0$ in all of these equations. This tells us immediately that

$$A = f(0), \quad B = f'(0),$$

$$C = \frac{1}{2}f''(0), \quad D = \frac{1}{2.3}f'''(0)$$

and so on.

In short,

$$f(x) = f(0) + xf'(0) + \frac{x^2}{1.2} f''(0) + \frac{x^3}{1.2.3} f'''(0) + \dots.$$

So, if we set aside some (thorny) questions such as convergence, the key to representing a function in this way is to know the values of the function and all its derivatives *at one particular point*, in this case $x = 0$.

97. Newton discovering 'Taylor series' in 1691/2, for the case in which $f(0) = 0$. He uses a dot to denote a derivative with respect to some fluxional variable t, as explained in Chapter 14.

The series is named after the English mathematician Brook Taylor, who published an equivalent result in 1715, but it seems to have been known to James Gregory as early as 1671. It can also be found—with just the same reasoning—in an unpublished manuscript by Newton of 1691/2 (see Figure 97).

The simplest example of a Taylor series in action is, perhaps, $f(x) = e^x$, because $f(x)$ and all its derivatives are then equal to 1 at $x = 0$, and the series reduces to the one we met in Chapter 22:

$$e^x = 1 + x + \frac{x^2}{1.2} + \frac{x^3}{1.2.3} + \ldots$$

But the functions $\sin x$ and $\cos x$ also lend themselves very easily to this kind of treatment, and repeated use of the results in Figure 70 leads to

$$\sin x = x - \frac{x^3}{1.2.3} + \frac{x^5}{1.2.3.4.5} - \ldots$$

$$\cos x = 1 - \frac{x^2}{1.2} + \frac{x^4}{1.2.3.4} - \ldots$$

And there is, in fact, a reason why I have placed these two series right next to the one for $e^x \ldots$

24

Calculus with Imaginary Numbers

In 1748, Euler took calculus in an altogether different direction with an extraordinary result linking e with the trigonometric functions (Figure 98).

$$e^{i\theta} = \cos\theta + i\sin\theta$$

98. A surprising connection.

The most remarkable feature here is the appearance of the imaginary number

$$i = \sqrt{-1},$$

which was, at the time, still treated with a certain amount of scepticism.

Yet, to see how this result comes about, all we have to do is take the infinite series for e^x from Chapter 23, pluck up a bit of nerve, and substitute in the imaginary number $x = i\theta$, where θ is real.

It is then just a matter of using $i^2 = -1$, over and over again, and collecting real and imaginary terms separately, to obtain

$$e^{i\theta} = \left(1 - \frac{\theta^2}{2} + \frac{\theta^4}{2\times3\times4} - \cdots\right)$$
$$+ i\left(\theta - \frac{\theta^3}{2\times3} + \frac{\theta^5}{2\times3\times4\times5} - \cdots\right),$$

whereupon the result follows, because the two series in brackets *are precisely those for* cos θ *and* sin θ in Chapter 23!

One particular case, obtained by setting $\theta = \pi$, is widely regarded as one of the most remarkable equations in the whole of mathematics:

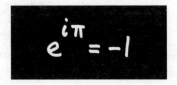

99. The most beautiful equation ever?

because it connects, in a most surprising way, the three special numbers e, i, and π. (Somewhat curiously, however, this particular equation never appears explicitly in any of Euler's writings.)

Functions of a complex variable

By around 1800 or so, the general idea of a *complex number*

$$z = x + iy,$$

where x and y are both real, was becoming more familiar, and mathematicians began to visualize such numbers, even, as points in a *complex plane* with real and imaginary axes (Figure 100).

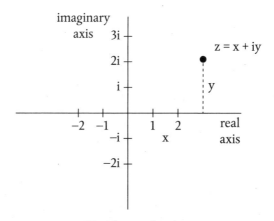

100. The complex plane.

And a little later still, in about 1820, the French mathematician Augustin-Louis Cauchy began developing the calculus of *functions* of a complex variable z.

Not surprisingly, this included the key idea of a derivative with respect to z, so that if w is some complex variable which is a function of z, then

$$\frac{dw}{dz} - \lim_{\delta z \to 0} \frac{\delta w}{\delta z}.$$

But while this definition might seem fairly innocent and innocuous, it isn't. This is because we could, in principle, take the limit $\delta z \to 0$ in many different ways, because—loosely speaking—we could approach the point z from many different directions in the complex plane. And requiring that all these different approaches give the same limiting value, dw/dz, depending only on the point z itself, turns out to have far-reaching and sometimes quite extraordinary consequences.

The calculus of flight

One of these consequences arose at the beginning of the 20th century, in the early days of aerodynamics.

The problem, in short, was: how can we find the airflow pattern around a wing?

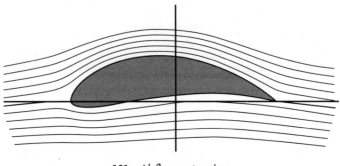

101.   Airflow past a wing.

In principle, of course, we write down, and solve, the appropriate differential equations of fluid motion.

In practice, however, the complicated shape of the wing, with its sharp trailing edge, poses severe mathematical difficulties (see Figure 101).

The corresponding problem for flow past a circular cylinder, on the other hand, is much easier to analyse mathematically (Figure 102).

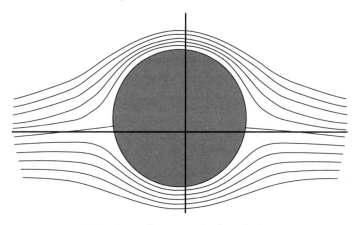

102. Streamlines past a circular cylinder.

And, quite remarkably, it is possible to apply a simple transformation to those streamlines, and get the flow pattern past the wing instead, if we view the streamlines as being *curves in the complex plane*.

In short, and however unlikely it might seem, it is possible to solve certain very real problems in fluid dynamics by leaping into the complex plane, applying a cunning transformation, and leaping out again.

25

Infinity Bites Back

Cauchy is mad…but, right now, he is the only one who
knows how mathematics should be done.

Norwegian mathematician Niels Abel, in a letter of 1826

During the 19th century, Cauchy and other mathematicians
fought to put calculus on a more rigorous foundation.

So much of the subject involved *infinity*, in one way or
another. And playing with infinity could be very dangerous
indeed.

A vanishing 'trick'

One example of this is the infinite series

$$1 - \frac{1}{2} + \frac{1}{3} - \frac{1}{4} + \frac{1}{5} - \frac{1}{6} + \cdots,$$

which is a bit like the Leibniz series, but uses all the whole
numbers instead of just the odd ones.

The series does, in fact, converge, and its sum turns out to
be $\log_e 2 = 0.693\ldots$

But suppose we now add up the terms *in a different order*, by taking two negative terms after each positive one:

$$\left(1 - \frac{1}{2}\right) - \frac{1}{4} + \left(\frac{1}{3} - \frac{1}{6}\right) - \frac{1}{8} + \left(\frac{1}{5} - \frac{1}{10}\right) - \frac{1}{12} + \cdots.$$

I am tempted to stress here that we are not 'missing out' any terms, or 'smuggling in' any new ones. Nor are we changing the sign of any of the terms.

It might seem, then, that the new series must inevitably converge to the same sum as before.

But it doesn't.

If we simplify all the brackets, we can rewrite the new series as

$$\frac{1}{2} - \frac{1}{4} + \frac{1}{6} - \frac{1}{8} + \frac{1}{10} - \frac{1}{12} + \cdots,$$

and this is equal to

$$\frac{1}{2}\left(1 - \frac{1}{2} + \frac{1}{3} - \frac{1}{4} + \frac{1}{5} - \frac{1}{6} + \cdots\right),$$

which is *half the sum of the original series*!

In other words, we seem to have made half of 0.693... 'disappear'.

Limits to the rescue

This 'vanishing trick' was discovered by Bernhard Riemann in 1854, and we can begin to understand it if we consider first two separate infinite series, one consisting of all the positive terms, and the other consisting of the negative ones:

$$1 + \frac{1}{3} + \frac{1}{5} + \frac{1}{7} + \cdots$$

and

$$-\frac{1}{2} - \frac{1}{4} - \frac{1}{6} - \frac{1}{8} - \cdots$$

And, as so often with infinite series, the safest way to proceed is to begin by considering the sum s_n of the first n terms, and *then* let $n \to \infty$.

The trouble is that, like one of the series in Chapter 9, neither of these series converges to a finite limit. In the first case $s_n \to \infty$ as $n \to \infty$, and in the second case $s_n \to -\infty$ as $n \to \infty$.

Suddenly, then, it is rather less surprising that if we combine the two the result will depend rather critically on how we do it.

Riemann went on to show, in fact, that we can make the resulting combination tend to *any limit we like* if we take the positive and negative terms in a suitably cunning order!

A Fourier series

A very different example comes from the series

$$y = \sin x + \frac{1}{3}\sin 3x + \frac{1}{5}\sin 5x + ...,$$

which arose in a study of heat conduction by Joseph Fourier, who was working in Paris in the 1820s.

It is, of course, unlike any infinite series that we have seen so far, because the individual terms are not simply powers of x.

Nonetheless, each individual term is a nice, continuous function of x, so it might seem reasonable to suppose that y will be also.

But it isn't.

If we plot y against x the result is a square wave, with y taking the value $\pi/4$ or $-\pi/4$ virtually everywhere, the only exceptions being wherever x is a multiple of π, where $y = 0$ (Figure 103).

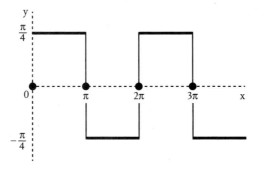

103. A 'square wave'.

In other words, the value of the function *y jumps* every time *x* passes through a multiple of π.

And yet, we again obtain some insight into what is happening if we consider the sum s_n of the first n terms. Some graphs of s_n against *x* are shown in Figure 104, and even on the basis of this small sample, it is possible to imagine how, at any particular fixed *x*, s_n tends to $-\pi/4$, 0, or $\pi/4$ as $n \to \infty$.

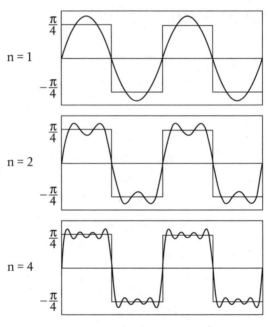

104. Graphs of s_n against *x*.

And just in case this is still not wholly convincing, it is worth noting that the result is certainly correct in the special case $x = \pi/2$, because it then reduces to the famous Leibniz series of Chapter 17:

$$1 - \frac{1}{3} + \frac{1}{5} - \frac{1}{7} + \cdots = \frac{\pi}{4}.$$

Limits everywhere

So limits are vital for a proper understanding of infinite series.

But *differentiation* is essentially a limit process, too, as we saw in Chapter 5:

$$\frac{dy}{dx} = \lim_{\delta x \to 0} \frac{\delta y}{\delta x}.$$

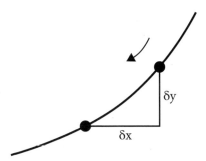

105. The derivative as a limit process.

And integration can also be viewed as a limiting process. Consider, for instance, the problem of finding the area under a curve between, say, $x = a$ and $x = b$, normally denoted by

$$\int_a^b y\,dx.$$

This whole problem started life, after all, with Fermat (and even—in a sense—Archimedes), as the limit of a sum (Figure 106).

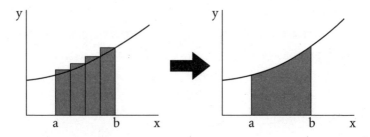

106. The integral as a limit process.

It is true that we have often viewed integration in this book as undoing differentiation, partly because that is what it often means in day-to-day mathematical and scientific practice, and partly because of the fundamental theorem of calculus (Figure 39).

But that theorem (and its proof in Chapter 8) is for when y is a *continuous* function of x, so that the curve has no gaps or jumps in it. And by the mid-19th century, mathematicians were beginning to take an interest in integrating far more general and badly behaved functions.

In any event, when Cauchy and Riemann set about trying to put calculus on a more rigorous footing, they both defined integration not as undoing differentiation, but as the limit of a sum.

On top of all this, some quite subtle questions were emerging about procedures involving *multiple* limits.

In Chapter 23, for instance, in order to find the coefficients $A, B, C, D \ldots$, we took an infinite series and differentiated it by differentiating each term.

But this amounts, in effect, to reversing the order of two limit processes (in that case '$n \rightarrow \infty$' and '$\delta x \rightarrow 0$'), and in general this is really quite risky.

Yet even as the 19th-century mathematicians began to grapple with matters as subtle as this, one crucial question remained.

What *is* a limit, exactly?

26

What *is* a Limit, Exactly?

I find it really surprising that Mr. Weierstrass...can attract so many students—between 15 and 20—to lectures that are so difficult and at such a high level.

colleague of Karl Weierstrass, 1875

What do we *really* mean when we say that

$$y \to 0 \quad \text{as} \quad x \to \infty,$$

or, equivalently, that the *limit* of y is 0 as x tends to infinity?

In order to explore this, I am going to suppose, in the first instance, that y is *always positive*.

And the first example which comes to mind is, perhaps, $y = 1/x$ (Figure 107). Yet, even with something as simple as this, why are we so sure that $y \to 0$ as $x \to \infty$?

The answer

   'y gets closer to 0 as x increases'

is plainly not good enough; the same could be said, for instance, of $y = 1 + 1/x$, which certainly doesn't tend to 0 as $x \to \infty$.

A better answer, surely, would be:

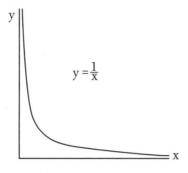

107. The function $y = 1/x$.

'We can make *y* *as close to* 0 *as we like* by taking *x* large enough'

and this is, I believe, the kind of thinking that we have used so far, from time to time, in this book.

But there is, in fact, still a problem.

The trouble is that definition would let in something like this:

$$y = \begin{cases} 1/x & \text{if } x \text{ is not a whole number} \\ 1 & \text{if } x \text{ is a whole number} \end{cases}.$$

This looks just like the graph in Figure 107 except that it has a sort of 'hiccup', and leaps up to the value 1, every time *x* is a whole number.

However artificial this example might seem, it is a perfectly valid function of *x*. It hardly conforms to any intuitive idea of '*y* → 0 as *x* → ∞', because it never completely 'settles down', but the awkward truth is that we *can* make *y* as close to 0 as we like by taking *x* large enough; we just have to be careful not to choose *x* as a whole number.

To kill off problems of this kind, then, we refine our definition further to:

> 'We can make y as close to 0 as we like for *all* x greater than some sufficiently large number.'

The only remaining difficulty then lies in the phrases 'as close as we like' and 'sufficiently large'. These are just too unwieldy for rigorous mathematical work, and so we refine our definition still further to:

> 'Given any positive number ε, there exists a positive number X such that $y < \varepsilon$ for all values of x which are greater than X.'

Behind this definition is the idea that it must work *no matter how small ε is*, but we don't need to state this explicitly, because it is covered by the key word 'any'.

And our simple example $y = 1/x$ does, indeed, conform to this, because, given any positive number ε, $y = 1/x$ will be less than ε for all values of x which are greater than $1/\varepsilon$.

Finally, however, we should relax our assumption that y is always positive, which I made purely for convenience of explanation. It might be, for instance, that $y \to 0$ in an oscillatory way as $x \to \infty$ (Figure 108).

As it happens, this final task is quite easy, for all we have to do is replace '$y < \varepsilon$' in our definition with '$-\varepsilon < y < \varepsilon$'.

This whole approach, which finally put the idea of a *limit* on a rigorous footing, is due to the German mathematician Karl Weierstrass, in the late 19th century.

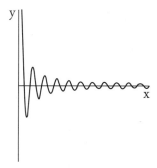

108. A decaying oscillation.

In retrospect, however, it is quite interesting to see just how close some earlier mathematicians came to the idea.

At one point in the *Principia* (1687), for instance, Newton wrote of one thing 'approaching' another

more closely than by any given difference.

And later, in 1765, D'Alembert wrote that one quantity is the limit of another

if the second approaches the first closer than any given quantity, however small.

With Weierstrass, however, all talk of 'approaching' has gone, to be replaced by extensive use of the inequality signs < and >. In this respect, it even contains faint echoes of Archimedes and Eudoxus, some 2000 years earlier.

Ultimately, however, Weierstrass's work looked forward, not back, and towards a more rigorous foundation for not only calculus, but even the whole idea of number itself.

And on that particular note I ought really to add what may seem a rather strange postscript.

Way back in Chapter 3, I said—quite correctly—that I do not know what it means for a number to be 'infinitely small'.

But there are mathematicians today who *do* deal with such numbers, in a branch of the subject called non-standard analysis, effectively started in the 1960s by the mathematician Abraham Robinson.

The truth, then, as I understand it, is that in order to be really certain about the foundations of calculus we eventually have to grapple successfully with *either* the idea of a limit *or* the idea of 'infinitely small'.

Until now, at least, most mathematicians have taken the first of these two routes, but in the end, it seems, the choice is ours.

27

The Equations of Nature

As we approach the end of this short book on calculus, I should like to return to its applications, and, first, to partial differential equations.

For these lie at the heart of so much of modern science, often in surprising ways.

Calculus and light

In 1865 the Scottish physicist James Clerk Maxwell formulated the mathematical theory of electromagnetism.

In particular, he discovered that electric and magnetic fields both satisfy the same partial differential equation.

And in its simplest form, in today's (S.I.) units, that equation is

$$\frac{\partial^2 y}{\partial t^2} = \frac{1}{\mu_o \varepsilon_o} \frac{\partial^2 y}{\partial x^2}.$$

Here, μ_o and ε_o are two electromagnetic constants which, even

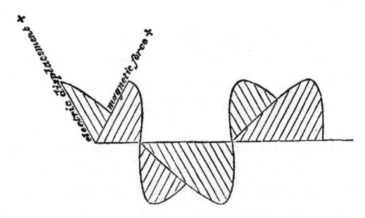

109. An electromagnetic wave, as sketched by Maxwell in his
Treatise on Electricity and Magnetism, 1873.

in Maxwell's time, were known, to some considerable accuracy,
from laboratory studies.

And if this equation strikes you as vaguely familiar, it will
be, I think, because, from a purely mathematical point of
view, it is *exactly the same equation as the one for a stretched string* in
Chapter 20!

The only difference is that in place of the constant T/ρ we
have a new constant, $1/\mu_o \varepsilon_o$.

Maxwell therefore knew immediately that the equation
would have wavelike solutions, and that these electromagnetic waves would travel with speed $1/\sqrt{\mu_o \varepsilon_o}$.

Moreover, this speed turned out to be so close to the measured speed of *light* that Maxwell at once concluded that light
itself must be an electromagnetic phenomenon.

And in this way, then, calculus played a major part in one of
the greatest discoveries in the history of science.

Calculus in the quantum world

Some 60 years later, in the 1920s, the world of physics was again in upheaval, this time with the advent of quantum mechanics.

This had been triggered, in part, by some experiments which could only be explained by regarding light not as a wave, but as a succession of particles called photons, each with a tiny amount of energy:

$$E = h\nu.$$

Here ν is the frequency of the light and h is Planck's constant $(6.626 \times 10^{-34} \text{ Joule sec})$.

Just as strangely, there were other experiments with particles—such as electrons—which could only be explained by viewing the particle as a wave.

To imagine all this, it can be helpful to think of a moving particle in quantum mechanics as a small packet of waves of limited extent (Figure 110).

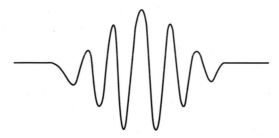

110.   A quantum wave packet.

And in 1926 Erwin Schrödinger introduced the idea of a *wave function*, ψ, to describe the form of a quantum mechanical wave, by writing down a differential equation for it.

In the simplest case, for a single particle of mass m moving in the x-direction in a potential V, Schrödinger's equation is

$$i\hbar\frac{\partial \psi}{\partial t} = -\frac{\hbar^2}{2m}\frac{\partial^2 \psi}{\partial x^2} + V\psi,$$

where $\hbar = h/2\pi$.

Once again, then, we find a partial differential equation at the heart of a physical theory, but this time with an interesting twist.

For the imaginary number

$$i = \sqrt{-1}$$

now appears directly in the differential equation itself. The wave function ψ is therefore complex, with real and imaginary parts that both depend on x and t.

Needless to say, then, this is all very different from the classical wave equation. And yet, one of the first successes of the full, three-dimensional Schrödinger equation was to account for the energy levels of an electron in a hydrogen atom:

$$E_N = -\frac{hcR_o}{N^2}.$$

Here c is the speed of light, R_o is Rydberg's constant $(1.097 \times 10^7 \, m^{-1})$, and, most importantly, $N = 1, 2, 3,...$ is a whole number.

The energy levels are therefore *quantized*, and if these discrete energy levels remind you at all of the discrete *frequencies* in Chapter 20, then you are in good company, because Schrödinger himself wrote:

> I wish to consider…the hydrogen atom, and show that… when integralness does appear, it arises in the same natural way as it does in the case of the node-numbers of a vibrating string.

Calculus goes supersonic

The 20th century saw major advances, too, in more classical areas of physics, including fluid dynamics. In particular, there was much excitement in the 1950s about the prospect of supersonic flight.

Even today most people know, I think, that something special happens when the speed of an aircraft passes through the speed of sound. But, from a mathematical point of view, *what is it?*

To answer this, it is simplest, I think, if we effectively move with the aircraft, so that the wing appears stationary.

Imagine, then, air moving with speed U, in the x-direction, past a thin wing. This will cause a small disturbance to the airstream, measured by a function of x and y called the velocity potential, ϕ. And it turns out that ϕ itself satisfies the partial differential equation

$$(1-M^2)\frac{\partial^2\phi}{\partial x^2}+\frac{\partial^2\phi}{\partial y^2}=0.$$

Here M is the *Mach number*, defined as

$$M=\frac{U}{c},$$

where c is the speed of sound.

It is immediately apparent, then, that as M increases past 1 the coefficient of the first term *changes sign*, and it is this change of sign which alters the whole character of the differential equation and, indeed, its solution.

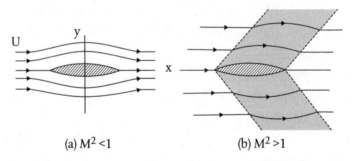

(a) $M^2<1$　　　　　(b) $M^2>1$

111. (a) Subsonic and (b) supersonic flow past a thin symmetrical wing.

For subsonic flow, with M < 1, the equation has strong links with the complex variable theory mentioned in Chapter 22, and there is some disturbance to the airstream everywhere, though it is very small at large distances from the wing (Figure 111a).

For supersonic flow, however, with M > 1, the equation becomes, essentially, the classical wave equation, and there is no disturbance at all to the airstream outside the two shaded regions in Figure 111b.

The *Mach lines* (dashed) that border those regions make an angle α with the x-axis such that

$$\sin\alpha = \frac{1}{M}.$$

So, the more supersonic the airstream, the smaller the value of α, and the more swept back the Mach lines.

Those lines themselves are, essentially, gentle versions of shock waves, and they travel along with the aircraft.

And, until the leading one arrives, a stationary observer on the ground hears absolutely nothing.

28

From Calculus to Chaos

Differential equations continue, to this day, to be the most important way in which calculus meets the real world.

And our ability to tackle them received an enormous boost in the 1960s, largely as a result of the computer revolution.

112. Chaos from the Lorenz equations: a path of a moving point with coordinates (x, z).

Calculus by computer

The basic idea is really quite simple, and dates back to the time of Euler.

Suppose we have a differential equation, such as

$$\frac{dy}{dt} = y.$$

As it happens, we know how to solve this particular one, from Chapter 22.

But suppose we didn't.

Imagine, instead, that we simply know the value of y—or at least a good approximation to it—at some particular time t.

Then the differential equation itself implies that a short time δt later the corresponding increase δy will be given, very nearly, by

$$\frac{\delta y}{\delta t} = y.$$

So, using our approximation to y at time t, we can calculate the small increase δy, add it to our 'current' value of y, and hence obtain an approximation to the 'new' value of y at time $t + \delta t$.

And, crucially, we can then take *that* value of y, and use exactly the same updating procedure to get an approximation to y a short time δt later still, and so on.

This whole approach is known as a step-by-step method, and should, in principle, give a good approximation to the true solution of the original differential equation *if we take a lot of very small time steps*.

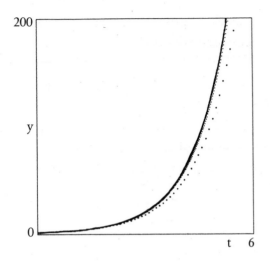

113. Euler's step-by-step method in action.

In Figure 113, for instance, we have tried to solve

$$\frac{dy}{dt} = y \quad \text{with} \quad y = 1 \text{ at } t = 0.$$

The lower 'curve' was obtained with $\delta t = 0.1$, and a gradual build-up of error is evident. The one above, however, was with the smaller value $\delta t = 0.02$, and is scarcely distinguishable on this time scale from the true solution $y = e^t$.

In practice, much more sophisticated and accurate ways of approximating δy are used.

Yet the basic idea is essentially the same—choose a small, fixed time step δt, replace the differential equation itself by an approximate updating procedure, and get a computer to implement that updating procedure, over and over again.

Most importantly, exactly the same idea can be used when dy/dt is given as some thoroughly awkward function of y.

And it can even be used if we have a whole *system* of differential equations like this involving several unknown variables.

Chaos

A famous example of this is provided by the *Lorenz equations*, which, in their most iconic form, are as follows:

$$\frac{dx}{dt} = 10(y - x)$$

$$\frac{dy}{dt} = 28x - y - zx$$

$$\frac{dz}{dt} = -\frac{8}{3}z + xy$$

They are therefore three differential equations for the three 'unknown' variables x, y, z as functions of time t.

A key feature of this system is that it is *nonlinear*. This is because of the terms $-zx$ and xy, which involve *products* of variables that we are trying to find. And it is that feature which makes these equations particularly challenging.

They first appeared in 1963 in a paper by the American meteorologist Ed Lorenz, where they arose from a highly oversimplified model of thermal convection in a layer of fluid.

Lorenz solved them using a step-by-step method on a very primitive 'desktop' computer, and if we do the same, and plot one of the variables against time t, we typically see oscillations.

But the oscillations are *chaotic*, and seemingly haphazard, so that the system never settles down to either a steady state or a regular, periodic motion (Figure 114).

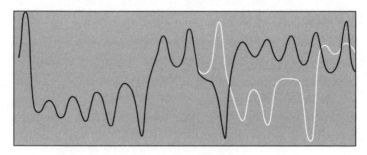

114. Chaos in the Lorenz equations, showing extreme sensitivity to initial conditions.

And there is a second, crucial, feature of chaos.

The black and white graphs in Figure 114 result from two initial conditions which are very slightly different, so that, at first, the two graphs are practically indistinguishable.

Yet, after just a few oscillations, the two graphs diverge substantially, and the system evolves thereafter in two completely different ways.

This extreme *sensitivity to initial conditions* is a key hallmark of chaos, and implies major problems with predicting the long-term behaviour of chaotic systems, simply because, in practice, the initial conditions may not be known to high accuracy at all.

This is a serious issue, for we now know chaos to be a common feature of many systems involving sets of nonlinear differential equations, whether in physics, engineering, chemistry, or biology.

And while some of the key ideas date back to the great French mathematician Henri Poincaré, in the late 19th century, the full importance of chaos came to be widely recognized only after the pioneering work by Ed Lorenz and others in the 1960s.

As it happens, Lorenz first came upon some of these ideas while working on an earlier, more elaborate, computer weather model involving *twelve* variables. And that model was, in turn, motivated in part by some remarkable laboratory experiments by the physicist Raymond Hide in the 1950s.

These involved a rotating water tank, of annular shape, with inner and outer boundaries maintained at different temperatures. In a sense, then, this was the atmosphere stripped down to its absolutely bare essentials: a basic uniform rotation and some differential heating (Figure 115).

(a)   (b)

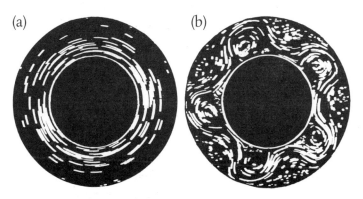

115. Two flows of a differentially-heated rotating fluid.

At low rotation speeds, the flow relative to the rotating tank was symmetric about the rotation axis (Figure 115a), while at higher rotation rates that flow became unstable, and a distinctive meandering flow structure (Figure 115b) emerged instead, reminiscent of the jet stream in the atmosphere.

But at higher rotation rates still, the wavy jet fluctuated in an irregular manner, and it was this behaviour that particularly intrigued Ed Lorenz.

And it was while Lorenz was trying to study a flow of this general kind, with his early 12-variable model, that fate lent something of a hand.

At some point, he decided to rerun a certain section of the output, so stopped the computer and typed in the initial conditions for that particular section. But, for practical reasons, he typed in not the original numbers, which were to 6-figure accuracy, but 3-figure approximations.

In his own words:

> I started the computer again and went for a cup of coffee.
> When I returned, about an hour later…I found that the
> new solution did not agree with the original one.

At first, Lorenz suspected some kind of computer failure, but he soon realised that the output itself told a quite different story.

For, while he had been having his coffee, the computer had simulated about two months of 'weather'. And, at first, the tiny round-off errors in the initial conditions made only small differences to the output.

Gradually, however, those differences steadily amplified, roughly doubling every four days or so, until sometime in the second month all resemblance to the original 'weather' completely disappeared.

It was in this way, then, that Lorenz stumbled, more or less by accident, on what we now call 'sensitivity to initial conditions', and he eventually came to the conclusion, even, that this extreme sensitivity is, to a large extent, the actual *cause* of chaos.

Ed Lorenz was a modest man, and saw himself, I think, as just one more scientist using mathematics—and particularly calculus—to try to understand how the world really works.

I once played tennis with him, in 1973.

FURTHER READING

From Calculus to Chaos by David Acheson, Oxford University Press, 1997.

A History of Pi by Petr Beckmann, Golem Press, 1970.

e—the Story of a Number by Eli Maor, Princeton University Press, 1994.

Calculus Gems by George F. Simmons, McGraw-Hill, 1992.

There are many scholarly books and articles on the history of calculus, but I particularly recommend:

Mathematical Thought from Ancient to Modern Times by Morris Kline, Oxford University Press, 1972.

Analysis by its History by E. Hairer and G. Wanner, Springer-Verlag, 1996.

The Historical Development of the Calculus by C. H. Edwards Jr., Springer-Verlag, 1979.

A full translation of Leibniz's 1684 paper can be found in *A Sourcebook in Mathematics 1200–1800* by D. J. Struik (Princeton University Press, 1986) or in *Mathematics Emerging; A Sourcebook 1540–1900* by Jacqueline Stedall (Oxford University Press, 2008), though I would direct any interested reader to *The Early Mathematical Manuscripts of Leibniz* by J. M. Child (Dover, 2005).

So far as Newton is concerned, I found Niccolo Guicciardini's *Isaac Newton on Mathematical Certainty and Method* (MIT Press, 2011) particularly helpful, together with D. T. Whiteside's monumental *The Mathematical Papers of Isaac Newton*, Vols 1–8 (Cambridge University Press, 2008).

REFERENCES FOR QUOTATIONS

Chapter 7
'I verily believe…'
The English Works of Thomas Hobbes of Malmesbury, Sir William
Molesworth, J. Bohn, 1845.

Chapter 8
'to understand this for sense…'
The English Works of Thomas Hobbes of Malmesbury, Sir William
Molesworth, J. Bohn, 1845.

Chapter 13
'…will be equal to $x.dy + y.dx$.…'
'Differentials, Higher-Order Differentials and the Derivative in the
Leibnizian Calculus' by H. J. M. Bos, in the *Archive for History of
Exact Sciences*, Vol. 14, p. 16, 1974.

Chapter 14
'In symbols one observes an advantage…'
A History of Mathematical Notations by F. Cajori, Open Court, 1929,
Vol. 2, p. 184.

Chapter 15
'They have changed the whole point…'
Isaac Newton on Mathematical Certainty and Method, referenced, by
N. Guicciardini, MIT Press, 2011, p. 373.

Chapter 17

'ye signes of ye series...'
The Correspondence of Isaac Newton, ed. H. W. Turnbull, Cambridge, 1960, Vol. 2, p. 181.

Chapter 26

'...more closely than by any given difference.'
Newton, I. (1687). *Philosophiæ Naturalis Principia Mathematica. Jussu Societatis Regiæ ac typis Josephi Streatii*. Londini.
'...if the second approaches the first...'
Analysis by its History, E. Hairer and G. Wanner, Springer, 1996, p. 171.

Chapter 27

'I wish to consider...'
Schrödinger, quoted in *An Introduction to Quantum Physics* by A. P. French and E. F. Taylor, Van Nostrand, 1978, p. 192.

Chapter 28

'I started the computer again...'
Ed Lorenz, quoted in *Bulletin*, Vol. 45, World Meteorological Organization, 1996.

PICTURE CREDITS

4. (a) Science & Society Picture Library/Getty Images. (b) Hulton Archive/Getty Images.

37. Wallis, John. *De sectionibus conicis nova methodo expositis tractatus.* The Bavarian State Library, 1655.

41. (b) I. Newton, *Analysis per quantitatum series, fluxiones, ac differentias: cum enumeration linearum tertii ordinis, Londini, Ex Officina Pearsoniana,* 1711, p. 19. Biblioteca Universitaria di Bologna (collocazione: A. IV.M.IX.28).

50. From Newton's Early papers, MC Add. 3958.4:78v. Reproduced by kind permission of the Syndics of Cambridge University Library.

53. Popperfoto/Getty Images.

56. Schooten, Franz van. (1657). *Exercitationum Mathematicorum.*

59. From Newton's Papers folio 56 of Add. 3965-7. Reproduced by kind permission of the Syndics of Cambridge University Library.

60. Leibniz, 1684, *Acta Eruditorum.*

67. From *Analysis by Its History,* E. Hairer and G. Wanner, p. 107, Springer, 1996. Reproduced by permission from Bibliothèque de Genève.

68. The Bodleian Libraries, The University of Oxford, call no. Savile ff8, *Analysis per quantitatum series, fluxiones, ac differentias* by Isaac Newton.

69. The British Library.

80. The Bancroft Library.

81. Bettmann/Getty Images.

89. Leibniz, 1684, *Acta Eruditorum.*

95. Culture Club/Getty Images.

96. From Newton's *De Analysi* (1669, published 1711) as it appears in *Analysis by Its History*, E. Hairer and G. Wanner, p. 54, Springer, 1996. Reproduced by permission of Bibliothèque de Genève.

97. Reproduced by kind permission of the Syndics of Cambridge University Library.

115. From R. Hide and P. J. Mason, *Advances in Physics*, Vol. 24, pp. 47–100, 1975.

INDEX

1089 AND ALL THAT

A Journey into Mathematics

David Acheson

DAVID ACHESON

1089
+ ALL THAT
= A JOURNEY INTO MATHEMATICS

An instant classic... an inspiring little masterpiece
Mathematical Association of America

978-0-19-959002-5 | Paperback | £8.99

'Every so often an author presents scientific ideas in a new way...Not a page passes without at least one intriguing insight...Anyone who is baffled by mathematics should buy it. My enthusiasm for it knows no bounds.' *Ian Stewart, New Scientist*

'An instant classic...an inspiring little masterpiece.' *Mathematical Association of America*

'Truly inspiring, and a great read.' *Mathematics Teaching*

This extraordinary little book makes mathematics accessible to everyone. From very simple beginnings Acheson takes us on a journey to some deep mathematical ideas. On the way, via Kepler and Newton, he explains what calculus really means, gives a brief history of pi, and introduces us to chaos theory and imaginary numbers. Every short chapter is packed with puzzles and illustrated by world famous cartoonists, making this is one of the most readable and imaginative books on mathematics ever written.

MATH HYSTERIA

Fun and games with mathematics

Ian Stewart

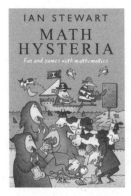

978-0-19-861336-7 | Paperback | £12.99

Professor Stewart presents us with a wealth of magical puzzles, each one spun around an amazing tale: Counting the Cattle of the Sun; The Great Drain Robbery; and Preposterous Piratical Predicaments; to name but a few. Along the way, we also meet many curious characters: in short, these stories are engaging, challenging, and lots of fun!

HOW TO CUT A CAKE

And other mathematical conundrums

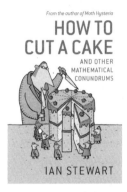

From the author of Math Hysteria

HOW TO CUT A CAKE
AND OTHER
MATHEMATICAL
CONUNDRUMS

IAN STEWART

978-0-19-920590-5 | Paperback | £11.99

Twenty curious puzzles and fantastical mathematical tales from Professor Ian Stewart, one of the world's most popular and accessible writers on mathematics.

Welcome to Ian Stewart's magical world of mathematics! This is a strange world of never-ending chess games, empires on the moon, furious fireflies, and, of course, disputes over how best to cut a cake. Each quirky tale presents a fascinating mathematical puzzle — challenging, fun, and also introducing the reader to a significant mathematical problem in an engaging and witty way.

COWS IN THE MAZE

And other mathematical explorations

Ian Stewart

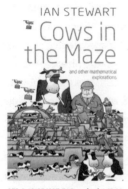

978-0-19-956207-7 | Paperback | £8.99

From the mathematics of mazes, to cones with a twist, and the amazing sphericon - and how to make one - Ian Stewart is back with more mathematical stories and puzzles that are as quirky as they are fascinating, and each from the cutting edge of the world of mathematics.

We find out about the mathematics of time travel, explore the shape of teardrops (which are not tear-drop shaped, but something much, much more strange!), dance with dodecahedra, and play the game of Hex, amongst many more strange and delightful mathematical diversions.

INFINITY

A Very Short Introduction

Ian Stewart

978-0-19-875523-4| Paperback | £7.99

The infinitely large (infinite) and the infinitely small (infinitesimal) are deeply fascinating topics, with connections to religion, philosophy, metaphysics, logic, and physics – and in mathematics many vital ideas – notably calculus – rest upon some version of infinity. Its history goes back to ancient times, with especially important contributions from Euclid, Aristotle, Eudoxus, and Archimedes. Cosmologists consider sweeping questions about whether space and time are infinite. Philosophers and mathematicians ranging from Zeno to Russell have posed numerous paradoxes about infinity and infinitesimals.

In this *Very Short Introduction*, Ian Stewart argues that working with infinity is not just an abstract, intellectual exercise but that it is instead a concept with important practical everyday applications, and considers how mathematicians use infinity and infinitesimals to answer questions or supply techniques that do not appear to involve the infinite.

NOTHING

A Very Short Introduction

Frank Close

978-0-19-922586-6 | Paperback | £7.99

'A fascinating subject covered by a fascinating book. - Marcus Chown, *Focus*

What is 'nothing'? What remains when you take all the matter away? Can empty space - a void - exist? This *Very Short Introduction* explores the science and the history of the elusive void: from Aristotle who insisted that the vacuum was impossible, via the theories of Newton and Einstein, to our very latest discoveries and why they can tell us extraordinary things about the cosmos.

Frank Close tells the story of how scientists have explored the elusive void, and the rich discoveries that they have made there. He takes the reader on a lively and accessible history through ancient ideas and cultural superstitions to the frontiers of current research.

FOUR LAWS THAT DRIVE
THE UNIVERSE

Peter Atkins

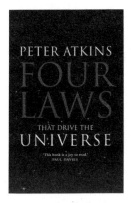

9780199232369 | Hardback | £13.99

'A brief and invigoratingly limpid guide to the laws of thermodynamics.' *The Guardian*

'Atkins's systematic foundations should go a long way towards easing confusion about the subject...an engaging book, just the right length (and depth) for an absorbing, informative read.' *Nature*

'Atkins' ultra-compact guide to thermodynamics is a wonderful book that I wish I had read at university.' *New Scientist*

The laws of thermodynamics drive everything that happens in the universe. From the sudden expansion of a cloud of gas to the cooling of hot metal, and from the unfurling of a leaf to the course of life itself - everything is directed and constrained by four simple laws. They establish fundamental concepts such as temperature and heat, and reveal the arrow of time and even the nature of energy itself.

Peter Atkins' powerful and compelling introduction explains what the laws are and how they work, using accessible language and virtually no mathematics.

EULER'S PIONEERING EQUATION

The most beautiful theorem in mathematics

Robin Wilson

The story of a supremely elegant equation which connects five of the most important concepts in mathematics

In just seven symbols, Euler's Equation connects five of the most important ideas in mathematics – our counting system; the concept of zero; the irrational number π; the exponential e; and the imaginary number i. Robin Wilson explains how mathematicians arrived at their understanding of each of these – and how Euler brought them all together.

978-0-19-879492-9 | Hardback | £14.99

CONJURING THE UNIVERSE

The Origins of the Laws of Nature

We know that the marvellous complexity of the Universe emerges from several deep laws and a handful of fundamental constants that fix its shape, scale, and destiny. The question Atkins addresses is How did these come into existence? They are, in Atkins's memorable words, the product of 'anarchy, indolence, and ignorance'.

Conjuring the Universe describes how laws such as the conservation of energy spring from deep symmetries, and explores how electromagnetism, thermodynamics, classical and quantum mechanics can all arise naturally out of the previous state – of absolute nothing.

978-0-19-881337-8 | Hardback | £14.99